Studies in Medieval History and Culture

Edited by
Francis G. Gentry
Professor of German
Pennsylvania State University

A Routledge Series

Studies in Medieval History and Culture

Francis G. Gentry, *General Editor*

1. "And Then the End Will Come"
Early Latin Christian Interpretations of the Opening of the Seven Seals
Douglas W. Lumsden

2. Topographies of Gender in Middle High German Arthurian Romance
Alexandra Sterling-Hellenbrand

3. Christian, Saracen and Genre in Medieval French Literature
Imagination and Cultural Interaction in the French Middle Ages
Lynn Tarte Ramey

4. Word Outward
Medieval Perspectives on the Entry into Language
Corey J. Marvin

5. Justice and the Social Context of Early Middle High German Literature
Robert G. Sullivan

6. Marriage Fictions in Old French Secular Narratives, 1170–1250
A Critical Re-evaluation of the Courtly Love Debate
Keith A. Nicklaus

7. Where Troubadours Were Bishops
The Occitania of Folc of Marseille (1150–1231)
N. M. Schulman

8. John Cassian and the Reading of Egyptian Monastic Culture
Steven D. Driver

9. Choosing Not to Marry
Women and Autonomy in the Katherine Group
Julie Hassel

10. Feminine Figurae
Representations of Gender in Religious Texts by Medieval German Women Writers
Rebecca L. R. Garber

11. Bodies of Pain
Suffering in the Works of Hartmann von Aue
Scott E. Pincikowski

12. The Literal Sense and the Gospel of John in Late Medieval Commentary and Literature
Mark Hazard

13. The Reproductive Unconscious in Late Medieval and Early Modern England
Jennifer Wynne Hellwarth

14. Mystical Language of Sense in the Later Middle Ages
Gordon Rudy

15. Fair and Varied Forms
Visual Textuality in Medieval Illustrated Manuscripts
Mary C. Olson

16. Queens in the Cult of the French Renaissance Monarchy
Public Law, Royal Ceremonial, and Political Discourse in the History of Regency Government, 1484–1610
Elizabeth A. McCartney

17. The Contested Theological Authority of Thomas Aquinas
The Controversies between Hervaeus Natalis and Durandus of St. Pourçain
Elizabeth Lowe

18. Body and Sacred Place in Medieval Europe, 1100–1389
Dawn Marie Hayes

19. Women of the Humiliati
A Lay Religious Order in Medieval Civic Life
Sally Mayall Brasher

20. Consuming Passions
The Uses of Cannibalism in Late Medieval and Early Modern Europe
Merrall Llewelyn Price

21. Literary Hybrids
Cross-dressing, Shapeshifting, and Indeterminacy in Medieval and Modern french Narrative
Erika E. Hess

The King's Two Maps

Cartography and Culture in Thirteenth-Century England

Daniel Birkholz

Routledge
New York & London

Published in 2004 by
Routledge
29 West 35th Street
New York, NY 10001
www.routledge-ny.com

Published in Great Britain by
Routledge
11 New Fetter Lane
London EC4P 4EE
www.routledge.co.uk

Routledge is an imprint of the Taylor & Francis Group

10 9 8 7 6 5 4 3

Library of Congress Cataloging-in-Publication Data
Birkholz, Daniel.
The king's two maps : cartography and culture in thirteenth-century England / by Daniel Birkholz.
p. cm. — (Studies in medieval history and culture ; v. 22)
Includes bibliographical references and index
ISBN 0-415-96791-0 (alk. paper)
1. Cartography—England—History—to 1500. 2. Cartography—Social aspects—England—To 1500.
I. Title. II. Series.
GA793.3.A1B57 2003
912'.0942'09022—dc21

2003014638

Series Editor Foreword

Far from providing just a musty whiff of yesteryear, research in Medieval Studies enters the new century as fresh and vigorous as never before. Scholars representing all disciplines and generations are consistently producing works of research of the highest caliber, utilizing new approaches and methodologies. Volumes in the Medieval History and Culture series will include studies on individual works and authors of Latin and vernacular literatures, historical personalities and events, theological and philosophical issues, and new critical approaches to medieval literature and culture.

Momentous changes have occurred in Medieval Studies in the past thirty years in teaching as well as in scholarship. Thus the goal of the Medieval History and Culture series is to enhance research in the field by providing an outlet for monographs by scholars in the early stages of their careers on all topics related to the broad scope of Medieval Studies, while at the same time pointing to and highlighting new directions that will shape and define scholarly discourse in the future.

Francis G. Gentry

Contents

List of Figures

Acknowledgments

I AM PLEASED TO THANK POMONA COLLEGE AND THE UNIVERSITY OF MINNESOTA Graduate School for research funding, travel grants, and other financial support making this study possible. Archivists and administrative staff at the institutions appearing in captions to my illustrations (too many to name individually) made my research trips in the UK and the U.S. both productive and enjoyable. Generous, timely assistance from librarians and reproduction specialists has enabled inclusion of the many figures on which my arguments are based.

The King's Two Maps had its genesis as an interdisciplinary project in a set of medieval seminars I had the good fortune to find at Carleton College; my thanks to Philip Niles, for his exemplary fire and rigor, and especially Sylvia Tomasch, for her wily provocation and ongoing professional support.

To train as a medievalist at the University of Minnesota in the middle 1990s was an exceptional experience. The atmosphere was one of cross-pollinated innovation: the faculty modeled a critical verve tempered by textual and historical precision. In conceptualizing, researching, and writing *The King's Two Maps* I benefited from feedback by Christian Aggeler, John Archer, Robin Brown, Barbara Hanawalt, Andrew Kincaid, Michal Kobialka, James Landman, Candace Lines, William Marvin, Lachlan Mead, Julia Mickenberg, Roger Miller, Andrew Morris, Colleen Page, Rita Raley, Jani Scandura, and Justin Steinberg, among others. Far above all, I am indebted to Rita Copeland and to David Wallace, whose mentorship has been as empathetic and as brilliantly generative as is their scholarship. I especially wish to thank them for the intellectual excitement they continue to inspire, and for the range, depth, and flexibility of their respective critical imaginations.

To my (now former) colleagues and students in and around the English Department at Pomona College, my heartfelt thanks for the past few years of reflective work and earnest play—I cannot imagine a community more enriching. Special appreciation to Martha Andresen, Toni Clark, Barbara Clonts, Edward Copeland, Kathleen Fitzpatrick, Rena Fraden, Gerald Frakes, Paul Mann, Cristanne Miller, Kevin Platt, Arden Reed, Paul Saint-Amour, Pamela Smith, Peter Stokes, Valorie Thomas, Margaret Waller, Gary Wilder, Bob Woods, and Steve Young.

To new English Department colleagues at the University of Texas at Austin—in particular Samuel Baker, Thomas Cable, Dolora Chapelle-Wojciehowski, James Garrison, Geraldine Heng, Liz Scala, and Marjorie Curry Woods—my gratitude for a welcome generously extended.

Though most of the work going into this book was conducted without the benefit of feedback from scholars specializing in early cartography, I have had the good fortune, more lately, to stumble into a number of cartographic circles, where I have met with uncommon warmth. My thanks, for their cartographic colleagueship (and/or their patient mentorship and methodological forbearance!) to Peter Barber, Surekha Davies, Paul Harvey, Alfred Hiatt, Scott Westrem, and especially Catherine Delano-Smith. My work has also benefited from the feedback provided by several anonymous readers. I should also like to acknowledge Mr. Kenneth Nebenzahl and Dr. James Akerman, for their sponsorship and administration, respectively, of the Nebenzahl Prize, awarded through the Hermon Dunlap Smith Center for the History of Cartography at the Newberry Library, Chicago, which *The King's Two Maps* (in earlier form) had the great honor to receive in 2001.

My closing thanks go to four family members: to Dean Birkholz and Janet Birkholz, for their perspective and unflagging belief; to Lena J. Mickenberg Birkholz, for laughter and inspiration; and finally to Julia Mickenberg, for debts and gifts too numerous and longstanding, too deep-running and pleasant to count.

Prologue

An Interest in Maps

In his *English Medieval Wall Painting* (1950), E. W. Tristam points out "an interest in maps on the part of our early kings." Tristam had immediately in mind a 1299 entry from the Wardrobe Accounts of Edward I, which lists *Unus pannus Regi datus ad modum mappe mundi* ("One measure of cloth given to the King with the form of a map of the world"). Almost nothing further is known about this particular "piece" or "scrap of cloth," except that it or another such *pannus depictus* ("painted cloth") is inventoried *in thesauraria apud Westmonasterium* ("in the treasuries at Westminster") again in 1302; and surprisingly, there has been little speculation.[1] But by no means is there a dearth of evidence linking English kings, particularly those of the thirteenth century, to maps of various kinds, particularly such large-scale, display-oriented "maps of the world," or *mappaemundi*, among which Edward's *pannus depictus* seems to number. In the pages to follow I will explore some implications of this link.

In doing so, I will revisit some assumptions that have become conventional among historians of cartography. To begin with I will underscore the limitations of a critical consensus that, assuming an operative ecclesiastical context for the entire genre, has tended to see medieval *mappaemundi* uniformly as "vehicles of religious instruction."[2] In response to this position, I will argue that a map such as one commissioned by King Henry III (1216–72) for the Painted Chamber at Westminster Palace in 1236 does more than mark its patron as pious. In addition, any map at Westminster—particularly one displayed in a chamber that saw some of the earliest meetings of Parliament—needs to be seen in the context of English and continental politics. Especially relevant are Henry's conflicts with his baronage, ongoing from the 1230s, and the related question, most intense in the 1250s, of the various crusade-inflected European ambitions of the royal house. Furthermore, despite the fact that the king's Painted

Chamber *mappamundi* does not itself survive (having been destroyed or painted over in the 1260s), such an image needs to be read in terms of its immediate physical context—its apparent place within an elaborate architectural and iconographic program which has been seen, on the whole, as an evolving illustration of Henry and Edward's developing views on the nature of English kingship.[3]

That a royal need for such arguments as a map of the world might uniquely provide was perceived to be ongoing or recurrent—and not limited to Henry and his mid-century struggles over the status of English monarchy—is suggested by the domestically secure but territorially ambitious Edward I (1272–1306) and his turn of the century "painted cloth." In an era marked by peripatetic aristocratic lifestyles and multiple royal dwellings and decorative sites, it seems telling that King Henry's map-mural and King Edward's (rolled or hanging) map-cloth each found a home at Westminster. It may be that Edward's "painted cloth" traveled with him, as a portable emblem or chamber dressing along the later lines of Richard II's Wilton Diptych or—to choose a controversial analogue from cartography itself—in the manner suggested by a triptych built to house the late thirteenth-century Hereford Cathedral *mappamundi* (fig. P.1).[4] Just as with Henry's Westminster mural, neither the geographical content of Edward's lost Wardrobe map, nor the precise social and material terms of its display, can be known for certain. Still, whatever the case—and keeping in mind that any answer we generate will require considerable speculation—surely it is worth asking what cultural work such images performed, on behalf of their (politically interested) royal patrons.

In order to understand the operative terms of any such cartographic "intervention" into thirteenth-century political affairs[5]—whether in the case of these two "King's Maps," or others—it will be necessary to reconsider, in addition to mainline interpretations of *mappaemundi*, a number of views commonly held concerning medieval cartography overall. The first of these is that medieval maps fall into "several quite distinct traditions" with little technical or ideological overlap; and the second, that medieval maps further separate into those "practical" and those "symbolic," categories to be seen in "fundamental contrast." The third assumption necessary to revisit is that any given medieval map bears a discernable and distinct "single function," beyond which interpretive attention is largely "irrelevant"; and the fourth is that maps from the various formal traditions, and of apparently contrasting functions or content-levels, are "best analyzed separately."[6]

There can be little doubting Henry III's personal familiarity with, even enthusiasm for, large-scale world map images. In addition to the Wardrobe Accounts' direct confirmation of his son's ownership of such images, various threads of indirect evidence and lines of scholarly argumentation tie Edward I, similarly, to prominent *mappaemundi*. What is more, compelling internal and external evidence links King Edward, through the medium of his enterprising royal clerks, to the cartographic document that is widely regarded as medieval England's most extraordinary: the Gough Map of Britain (fig. P.2). (The Gough Map's extant copy dates to c.1360, but its prototype appears to have been developed beginning about 1280, with this template afterward being augmented and updated all the way until the mid-sixteenth century.)[7]

Map historians have tended to view medieval cartography's several discernable genres in categorical isolation from one another. I will propose, however, that rather than exemplifying "fundamental contrast" in cartographic form and social function, maps of the two major types cited above—maps of the world, like King Henry's mural or King Edward's *pannus depictus*, and maps of the realm, like the Gough Map with its suggestive links to royal clerks—not only *can* be read together but on a hermeneutic level *need* to be, insofar as they display significant formal and functional alignment. Far from possessing negligible overlap in terms of either their construction or their cultural meaning (such as would render these genres "best analyzed separately"), strong case can be made that the latter form—the freestanding/isolated map of the realm or developing secular state, unknown before about 1250—effectively emerges, indeed grows inexorably out from that established model for imagining and organizing space which (as we shall see) is so provocatively incarnated in the world map. Moving from attempted reconstruction of the formal features (and social meanings) of Henry III's Westminster *mappamundi*, to close examination of the Gough Map, in the context of Edward I's politics of insular British conquest, I will set out the implications of this important but unrecognized development in cartographic history for English cultural and political history more generally. What I will suggest, in specific, is that early examples of these two crucial premodern cartographic genres—maps of the world and of the realm—effectively combine to figure (that is, help establish visual and conceptual terms for) a specifically royalist ideology linking English kingship with territorial statehood, both of which categories were under intense negotiation in the period.

I take my title in this study—*The King's Two Maps*—not from any particular pair of extant medieval documents so much as from the larger phenomena of these two map-genres—*mappaemundi* and, to coin a term, *mappaeregni*—that are traceable to thirteenth-century English royal interest. When seen together, maps of the world and maps of the realm help elaborate, on Edward I's behalf, a vision of kingship and territoriality that is both sacred and imperial, at once centripetal (retrenching inwards, upon a symbolic and bureaucratic Westminster core) and centrifugal (expanding outwards, in acquisitive territorial desire). To put it another way, if read in tandem *mappaemundi* and *mappaeregni* help underline cartography's importance, as an emergent and increasingly "official" discourse, to key developments in both "medieval political theology" and early English governmental practice.[8] We shall see that maps play a decisive role, first, in the articulation and representation of a system of centralized legal and financial (or "fiscal") administration that would accrue important momentum in the late thirteenth century; and second, in practices of ideological and military domination, by a nascent territorial state in a period usually seen as previous to and therefore largely innocent of these practices. Maps from later periods have been understood for some time as being not just implicated in but integral to processes of political domination—"especially in periods of colonial history." Likewise have Roman maps been interpreted, both as metaphors of dominion and as more directly practical participants in programs of military conquest and provincial administration.[9] But in part, I think, because commentary on medieval maps of the world—the period's most spectacular as well as most studied cartographic form—has not stressed enough the key involvement of secular governmental interests in their commission and presentation (if not indeed their production), and has embraced instead a model which sees the world map in relative generic isolation, primarily as a kind of politically disinterested, salvation-obsessed, "preaching" or "teaching" religious tract, a resultingly oversimplified, over-theologized vision of not only *mappaemundi* themselves, but of medieval cartography and indeed medieval spatiality as a whole, has emerged.[10]

The present study builds, indispensably, on the secure foundation laid by previous workers in the field of historical cartography—those who have produced a scholarship of painstaking, minute description; of authorship, patronage, and provenance; of formalist comparison and geneaology; of materialist inquiry into artifact construction and setting; of technical geographical analysis; of map history at key moments as well as over the *long durée*; and more. In a mode potentially at odds with established practice in the subdiscipline now known as the History of Cartography, however, *The King's Two Maps* draws simultaneously on a tradition of historicist analysis that is not only speculative (when called for) in the face of a laconic material record, but is moreover frankly ideological in certain of its dimensions: in the awkward questions it asks of art-objects, in the connections it insists upon drawing between cultural expression and political interest. Taking as its point of departure, in this vein, J. B. Harley's proposition that historically maps (even more so than other forms) have been "preeminently a language of power, not of protest,"[11] the narrative to follow will suggest how an existing model of church cartography was in the thirteenth century seized upon, by those looking to promote the interests of Henry III and Edward I, and its visual language adapted for use in some of the most important, and disputed, political projects of the day.

PATRONAGE, PART ONE: CHURCH CARTOGRAPHY

Harley has taken pains to point out the close association of maps, in almost every period, "with elite or powerful groups or individuals." Elsewhere—in their landmark *History of Cartography Volume I* (1987)—he and David Woodward have stressed that "the links between the character of particular elites, the institutions through which their power was exercised, and the types of maps they produced" deserve particular attention. "Questions of social and cultural context," Harley and Woodward maintain, "go to the root of understanding in the history of cartography."[12]

In this light, it bears noting that for some years now medieval *mappaemundi* have, as a genre, been understood primarily as "the emanations of power of a clerical elite." To be certain, a majority of *mappaemundi* did exist within an ecclesiastical context: in monastic libraries. However, in the main, these were neither freestanding nor architectural icons, productions embedded within monumental pictorial programs or meant for portable display on their own. Instead, most world maps were bound within manuscripts as illustrations or diagrams accompanying texts including Beatus' *Commentary on the Apocalypse of St. John*, Orosius' *Historia adversum paganos*, Isidore of Seville's *Etymologiae* and *De natura rerum*, Honorius of Autun's *Imago Mundi*, Ranulph Higden's *Polychronicon*, and others. This distinction between a textual and diagrammatic as opposed to a monumental display orientation is crucial. Inattention to it has meant that the literature of the history of cartography has slid too often into the assumption—largely unfounded—that large-scale *mappaemundi* were, like codices containing world map drawings or miniatures, "probably found in a number of the great religious houses of Europe in the Middle Ages."[13]

Evidence does exist of an ecclesiastical display context for some large *mappaemundi*, such as a wall-map made for Evesham Abbey in about 1390. Other English examples now lost (i.e., productions known or believed to have existed, but no longer extant) include one present at Lincoln Cathedral in the mid-twelfth century; another bequeathed to Durham Cathedral Priory in the late twelfth century; and a third displayed at Waltham Abbey (Essex) into the mid-thirteenth. Moreover, the provenance of two recently discovered medieval map fragments indicates that large-scale *mappaemundi* likely also were present at Creake Abbey (Norfolk) during the fourteenth century and at a College of the Bonhommes founded in 1283 at Ashridge (Hertfordshire) by Edmund, Earl of Cornwall—first cousin of King Edward I. Provocative work by Marcia Kupfer, on a lost mural *mappamundi* in the nave at Chalivoy-Milon near Bourges in central France, and by Ernst Kitzinger, on a partially preserved cartographic floor mosaic from the site of what is now Turin Cathedral, has helped establish—at least for the continent and for the middle twelfth century—the presence of a fully fledged "tradition" of monumental world maps "commissioned for public display in ecclesiastical contexts." Finally, there are the two most celebrated large-scale *mappaemundi* to have survived from the thirteenth century to the twentieth, those held at Hereford Cathedral (fig. P.3) and at the convent of Ebsdorf in Lower Saxony (fig. P.6, below). Both of these—at least until quite recently—have been interpreted predominantly within the terms of this "ecclesiastical" "tradition"; they tend to be cited, that is, as paradigms of "church cartography." An eighteenth-century suggestion that the Hereford *mappamundi* "served anciently as an altarpiece in this church" has, for example, enjoyed considerable currency ever since.[14]

Backed by such material, it was with some confidence that Harley and Woodward reported in 1987 that an identifiable overall "meaning" for medieval *mappaemundi*—both as a coherent genre and "within their wider cultural and societal context"—was "starting to emerge": "The primary purpose of these *mappaemundi*," they said, "was to instruct the faithful about the significant events of Christian history." Woodward and others have since sounded occasional cautionary notes ("we have to be careful about assigning only a spiritual meaning to *mappaemundi*"), or have offered limited qualifications to account for variation from type. But in effect there has come to be near unanimity among historians of cartography in support of interpretations that highlight the "theological function" of the medieval world map.[15]

Maps like Hereford and Ebsdorf have been described as Roman documents adapted to fit Christian doctrine. The centering of many *mappaemundi* on Jerusalem, for example—a common but not universal practice—is a feature that can claim biblical justification, and even a rough "geographical" one, but this feature owes a direct formal (not to mention ideological) debt, more importantly, to the late-antique practice of centering imperial maps on Rome, out from which routes to the provinces were sometimes shown as radiating (fig. P.4). Medieval world maps of the well-known 'T-in-O' type (common but by no means exclusive as a format for *mappaemundi*) also derive their basic geometric structure from classical models. In this scheme the Mediterranean Sea combines with the River Nile (extending south) and the River Don (extending north) to sketch a watery "T" that divides the earth's circular land-mass into Europe, Africa, and Asia, while an "O" is drawn by an encircling ocean (fig. P.5).[16]

Exactly how medieval "church cartography" distinguishes itself from the formal practices (and effective meanings) of its classical forbears can be seen most dramatically in the example—admittedly exceptional—of the Ebsdorf Map (fig. P.6). Along with the miniature London Psalter Maps we shall examine in chapter one, Ebsdorf constitutes one of the medieval period's most arrestingly macrocosmic cartographic images. It is also one of a number of compositions, in variations to be found across the spectrum of medieval illuminated books, which share a loose thematic affiliation. I do not suggest that these images form a coherent group or visual genre in any conventional sense, nor do I refer primarily to maps per se; their common element is that they blend, juxtapose, or otherwise bring together a bodily representation of the divine—typically God as world's creator and/or Christ as world's redeemer—with a more or less map-like disc, a T-O orb, or some other spherical or circular figure representative of incarnate geography (sometimes accompanied or replaced by natural flora and fauna, or astronomical icons).

Potentially overwhelming as an amalgam of word and image, the Ebsdorf Map brings a unifying conceptual and geometric coherence to its dense encyclopedic diversity by superimposing the circular and tripartite "known world" (as imagined by medieval Europeans) onto the crucified body of Christ. Christ's head, hands, and feet figure strikingly as poles on the map, set in place as it were of the cardinal directions (fig. P.8). And like disembodied, almost reliquary emblems, they seem to operate as stations of geographical devotion. *Mappaemundi* placed Jerusalem at the center of their own textual/geographic body with increasing frequency in the years following the First Crusade (1095–99), making of it the very "navel of the world" and the point in the composition which commanded visual attention most powerfully. Moreover, the alignment of a symbolic Christian geometry (the cross) with the abstract geography invoked by reference to the four cardinal directions reinforces the form's structurally implicit argument that the entire created world is a manifestation of God's power—is in essence the incarnate and communal body of Christ. This "traditional association" between the four cardinal directions and the icon of the cross (as Woodward has observed) is seen also in the cruciform plan of churches, which like *mappaemundi* were typically oriented to the east.[17]

Unlike Ebsdorf, in the elaboration of its visual exegesis the Hereford Map chooses not to textualize so explicitly the association of the body of Christ with the circular mass of the created world. However, this map places the totemic scene of the crucifixion—a sketch dominated by its own starkly linear "T"—directly above the (circular) city of Jerusalem. Moreover, Jerusalem itself sits just above the cross of the larger, schematic T's river-arms with its vertical Mediterranean stem (fig. P.9). Scholars differ on the question of whether they discern a (typological, if not literal) "T" lying beneath the system of revealed waterways giving convoluted outline to the world map's familiar three continents. Yet inasmuch as the "T" in the T-in-O schemata almost certainly represented a cross—not initially, but according to the exegetical thinking characteristic of the scholastic Middle Ages—it would appear that allusion to Christ's passion is to some degree inherent in the formal and hermeneutical structure of thirteenth-century *mappaemundi* (those of this subtype, at least).[18] As the Hereford Map's underlying geometry and central crucifixion sketch help emphasize, such a conclusion may apply even to maps which did not follow the

Ebsdorf Map's practice in incorporating Christ's body explicitly into the disc of the world.

Modern commentary has dwelt (until recently) largely on the didactic implications of some *mappaemundi*'s religious or academic contexts (devotional, institutional, exegetical); on the form's visual rehearsal of the stages of Christian universal history (i.e., how medieval cartography maps time as well as space); and on specific content-level details which serve further to underline an essentially religious reading of the genre, such as the Hereford Map's illustration of a pelican piercing its own breast to feed its young, traditionally a symbol for Christ.

To be sure, theological interpretations have much force, and do seem to describe a "primary purpose" or dominant social function of many maps, extant or known. As we have noted, however, not all large-scale *mappaemundi* may be described as having been "commissioned for public display" in out-and-out "ecclesiastical contexts"—whether or not we include within this rubric the category of the devotional, so crucial to any discussion of medieval art patronage. Even a phantom production such as Edward I's Westminster map-*pannus*, though we know nothing beyond its 1299/1302 enrollment in the Wardrobe Accounts, prohibits confidence in a uniformly ecclesiastical or devotional presentation-setting (and hence interpretive context) for *mappaemundi*. Medieval English documents relating to art regularly pair the Latin *pannus* with *altaria* (n.pl. "altar, high altar"), as in *pannus ad altare*, *pannus ante altare*, *pannus pro altari*, *pannus super altare*, and so forth ("a measure of cloth at/in front of/before/above the altar"). Negative evidence is of course not often conclusive; assuming for example that Edward's two Wardrobe scraps *ad modum mappe mundi* are one and the same, it is notable that in 1299 this object is described *pannus datus Regi* with no mention of its being "painted," while in 1302 it is *pannus depictus* with no mention of its "having been given to the king." Still, were this royal-owned *mappamundi* designed as a portable altarpiece, or regularly used in chapel devotions during these years, we may wonder whether it would not have been described like others of the sort, that is, as perhaps *pannus ad modum mappe mundi ante altare*—there being precedent for the construction *pannus . . . ante altare*.[19]

Despite a distinguished history of assumption to the contrary, it is similarly far from clear that Hereford and Ebsdorf served originally (or at any time in medieval years) as altarpieces—or more generally, even as sanctuary decoration, whether being designed for or adapted to this use. Earlier we began to emphasize the role played in medieval map production by intellectual elites, "the few literati associated with [or drawn from] the ruling classes"—which is to say, those exceptionally learned churchmen who designed, researched, and apparently directed the execution of (if they did not themselves actually draw, paint, or letter) *mappaemundi* along monumental lines like these. However, no matter who compiled and executed them, because they were by all estimates extraordinarily expensive, large-scale maps above all else "had to be 'commercial,' that is, pleasing to the customer."[20] Much as for the composite category of medieval textual "authorship," in which responsibility is shared among various agents (scribe, compiler, patron, *auctor*), the dynamics of artistic production in thirteenth-century England do not typically allow for clean and confident division. This is true for questions of artistic impetus and contribution, but even more the case when it comes to attempted delineation between the interpenetrating realms of the secular and ecclesiastical. In the interest, therefore, of charting additional meanings for medieval world maps—beyond but *necessarily inclusive of* the theological, not dispensing with but building upon meanings that have been developed by scholars already—it seems useful to consider the potential investment in their display shown by that second order of "customer": the royal patron, whose ideological interests often coalesce with the church's, but just as frequently strike out in directions all their own.

PATRONAGE, PART TWO: ROYAL CARTOGRAPHY

Direct or indirect royal patronage has been established, or can be suggested, for many of medieval England's major known *mappaemundi*. In addition to those mentioned already, in 1239 Henry III ordered (as his Liberate Rolls testify) "a map of the world to be painted in the hall"—the recently completed Great Hall—of his palace at Winchester. More central to this study, however, is an inscription that appears on a world map sketch by the Benedictine chronicler and illustrator Matthew Paris, found in a manuscript of c.1250–55. Regarding the comparatively rough map it accompanies (Richard Vaughan observes that the artist "does not seem to have thought very highly" of his model), Matthew's comment explains that "this is a reduced copy of the world maps of Master Robert Melkeley and Waltham [Abbey]." More importantly—and somewhat oddly—Matthew's inscription next goes on to offer the (presumably related) information that elsewhere he has "most accurately copied" *mappa mundi dominis regis quod est in camera sua apud Westmonasterium*: that is, "the Lord King's map of the world which is in his chamber at Westminster." Here, in the Palace's *camera depicta* or "painted chamber," a prominent large-scale world map seems to have been present from 1236 until a 1262/63 fire destroyed all or part of it (along, apparently, with the larger mural cycle of which it formed an integral part). If medieval *mappaemundi* were most appropriately altarpieces—or are to be classed primarily as one form or another of sanctuary decoration—it may be worth asking why Henry chose such images for his Winchester Castle Great Hall and Westminster Palace bed-and-audience-chamber, instead of for, say, Winchester *Cathedral* and Westminster *Abbey*, sites upon which he also lavished a good deal of decorative attention throughout his reign.[21]

A basic association between maps and secular rulers was, even to the England of the thirteenth century, itself nothing new. Charlemagne's biographer Einhard mentions three silver tables, engraved with maps of Constantinople, Rome, and "the whole world in three concentric circles," among the items in the Carolingian emperor's will. Significant Roman precedents include Julius Caesar's well-known imperial project, a survey of the known world begun in 44 BCE; the Antonine Itinerary, a compilation of routes in the Roman provinces prepared for one of the Antonine emperors, probably Caracalla (CE 214/15); and most notably an immense world map compiled by Marcus Vipsanius Agrippa (63–12 BCE) at the request of his father-in-law, the emperor Augustus, who after Agrippa's death oversaw completion of the work himself and had it displayed on a colonnade at Rome (c.7–2 BCE). The cartographic productions here in question differ as much in material and formal specifications, clearly, as they do in terms of their political, religious, and cultural contexts. Nonetheless, as years of scholarship in historical cartography indicate, an operative link between technologies of mapping and the needs of secular rulers (practical or symbolic) not only obtains long beforehand but continues well beyond the medieval period.[22]

Scholars of medieval cartography have frequently invoked the concept of an especially fertile "English geographical culture of the thirteenth century," but on the whole stress ties that are ecclesiastical in their discussion of large-scale *mappaemundi* as a characteristically English genre. Peter Barber's voice has emerged as one that will attest to the decisively secular orientation of many premodern maps. Barber suggests that it is not until the middle of the sixteenth century that maps were systematically "enrolled into the service of English government" as "a fully integrated aid in the formulation of policy and an instrument of administration." Our later discussion concerning the Gough Map of Britain, in chapter three and parts of chapter two, may look to nudge this judgment backward to account for similar use as early as the reign of Edward I. But what is more, even such thirteenth-century *mappaemundi* as are only tenuously associated with England are striking in that they owe what links they have to royal contact, in particular. That is, what is arguably more important about (for example) the early thirteenth-century Vercelli Map, than that "its inspiration [if not layout] may well have been English," is that the man who apparently brought it back to Italy, Cardinal Guala Bicchieri, served until 1218/19 as Papal Legate to Henry III. In 1216 Guala Bicchieri had presided—not so incidentally, as we shall see later—at a young Henry III's coronation ceremony.[23]

Gervase of Tilbury, long regarded as a possible compiler of (or inspiration for) the 1235–39 Ebsdorf Map, has also been connected with the royal house of England. Gervase's presence has moreover been noted (on documentary grounds) at the court of Henry III in 1229—just prior to a decade that saw not only completion of this most elaborate of known medieval maps, but also King Henry's commission of two comparably scaled *mappaemundi*, to adorn his palaces at Winchester and Westminster. In a compelling study of the place-names included within the bounds of the Holy Roman Empire on the Ebsdorf *mappamundi*, Armin Wolf has argued that while it may have been made and preserved at Ebsdorf, nonetheless this monumental image most likely was intended for display in the ducal residence at nearby Lüneberg of Otto the Child, Duke of Brunswick (1213–1252)—to whose uncle (Emperor Otto IV of Brunswick, nephew of Richard I of England) Gervase had dedicated his *Otia imperialia* in 1214/15, and on whose behalf he had come to Westminster in 1229. The documentary evidence may never be conclusive, yet it seems reasonable to connect Gervase and his diplomatic visit, at least provisionally, with the interest in cartographic iconography that was to develop in Henry in the 1230s. A relevant, indirectly related point may reside in a curious detail of toponymy (i.e., the social geography implied by whatever set of place-names appears on a given map). For the cartographic sketch by Matthew Paris referred to above—the one containing gratuitous mention of the king's Westminster *mappamundi*, with emphasis on how Matthew has made careful figure or copy of it—apparently displays, in its particular roster of toponyms, considerable correspondence with Ebsdorf's. In addition, then, to the impetus provided by any native insular "geographical culture," it would appear that the phenomenon of thirteenth-century cultural "internationalism" played its part in the development of English world map iconography.[24]

In fact, according to the suggestion of specialists in cartographic provenance and genealogy, Henry's mid-thirteenth-century Westminster mural figures as a crux in the stemma of medieval English *mappaemundi*. The crucial manuscript inscription explains that the Lord King's map of the world *figuratur in ordine Matthaei de Parisio*—"is most accurately copied in the ordinal of Matthew of Paris"—but unfortunately, this book appears no longer to exist. On the other hand, recent scholarship by Peter Barber and by Graham Haslam has suggested (the proposals come independently) that Henry III's lost Westminster *mappamundi* may have served as a prototype, either directly or through an intermediary such as Matthew's lost ordinal copy, not only for the two well-known miniature "Psalter Maps" of c.1260–65 but also for the two map fragments of English provenance mentioned above, the Aslake Map of the middle fourteenth century (linked to Creake Abbey) and the Duchy of Cornwall Map of c.1260–85 (associated with the founding of Ashridge College). Similarities between this "newly recognized stream of cartography" and other genealogical streams (i.e., *mappamundi* sub-genres) associated with the Hereford Map and the Ebsdorf Map, respectively, have also been accounted for through recourse to King

Henry's lost map. In addition to Gervase of Tilbury's in 1229, a visit to Westminster has been posited, in the early 1260s, for Richard de Bello, a proposed alias of Richard of Haldingham whom some (though fewer and fewer with much confidence) regard as a candidate for compiler of the Hereford Map.[25]

The point I wish to make, very simply, is that a number of maps appear to derive from or share features with an exemplar commissioned by King Henry III himself. This Westminster map-mural amounts to an exemplar of the highest possible profile, having been produced at major expense (then augmented with a growing roster of flanking images) and what is more apparently designed to direct royal specifications. Finally, this image was set in a location (Henry's Painted Chamber) remarkable not merely on account of its visuals, but in terms of the daily performances of theatrical monarchy it saw; for this bed-and-audience-chamber (so we shall see) served as a primary point of access to the royal body, hence to all manner of social, political, and cultural influence. In sum, Henry's map was placed just where an extravagant image would be sure to make a strong and widely disseminated impression.

A rich documentary and material record testifies to Henry III's deeply felt and highly specific ideas about art, and especially the iconography of "pious" or sacred kingship.[26] Extant works and the documentary record describe Edward I as a patron of art and architecture less fully, but his martial tastes and close attention to the legal, financial, and administrative aspects of rulership emerge clearly nonetheless. To the extent that there are formal differences but at the same time thematic and graphic congruences between earlier and later thirteenth-century English royal maps, then—between *mappaemundi* associated with Henry and *mappaeregni* (or regnal maps) tied in earliest form to Edward—it may be that we can see in such transference the evolution of royal ideas about the general political usefulness, and precise policy implications, of cartographic imagery.

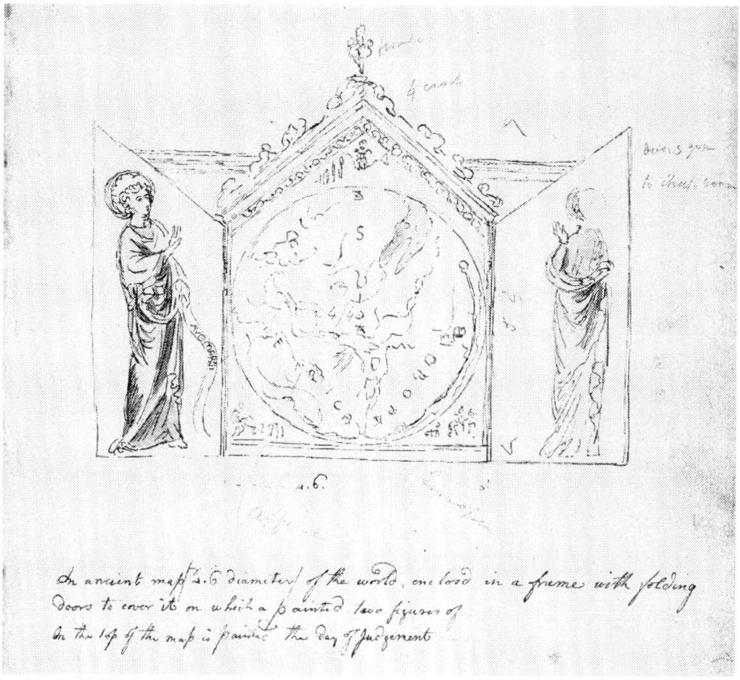

Figure P.1. Wooden triptych containing the Hereford *Mappamundi*; drawing by John Carter, c.1770s; London, B.L. MS Additional 29942, f.148; reproduced by permission of the British Library.

Figure P.2. The Gough Map of Britain, c.1280–1360; Oxford, Bodleian Library Gough Gen.Top.16; see fig. 3.1, below, for facsimile version; reproduced by permission of the Bodleian Library.

Figure P.3. The Hereford Cathedral *Mappamundi*, c.1285–1300; reproduced by permission of the Dean and Chapter of Hereford and the Hereford Mappa Mundi Trust.

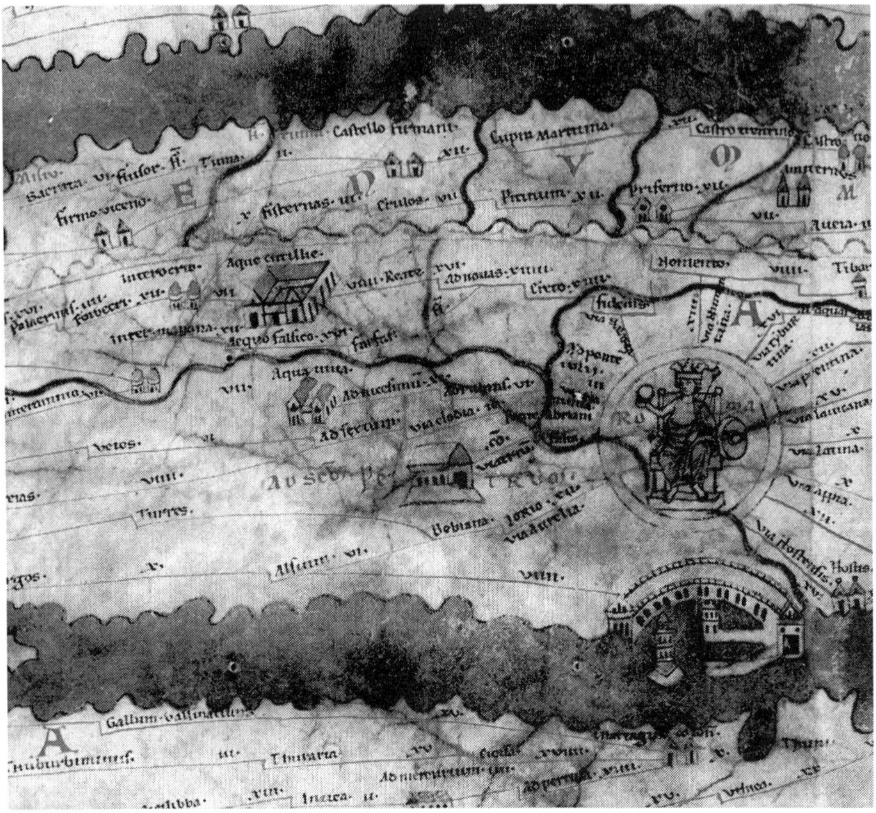

Figure P.4. The Peutinger Table: Rome; Vienna, Österreichische Nationalbibliothek, Codex Vindobonensis 324, Segment IV (detail); copied c.1230, late-antique prototype; note the City of Rome's personification as an armed and enthroned goddess, holding globe or orb; reproduced by permission of the Bildarchiv d. ÖNB, Wien.

Figure P.5. T-O Map Schema, from a printed copy of Isidore of Seville's *Etymologiarum* (Augsburg, 1472); reproduced by permission of the Newberry Library, Chicago.

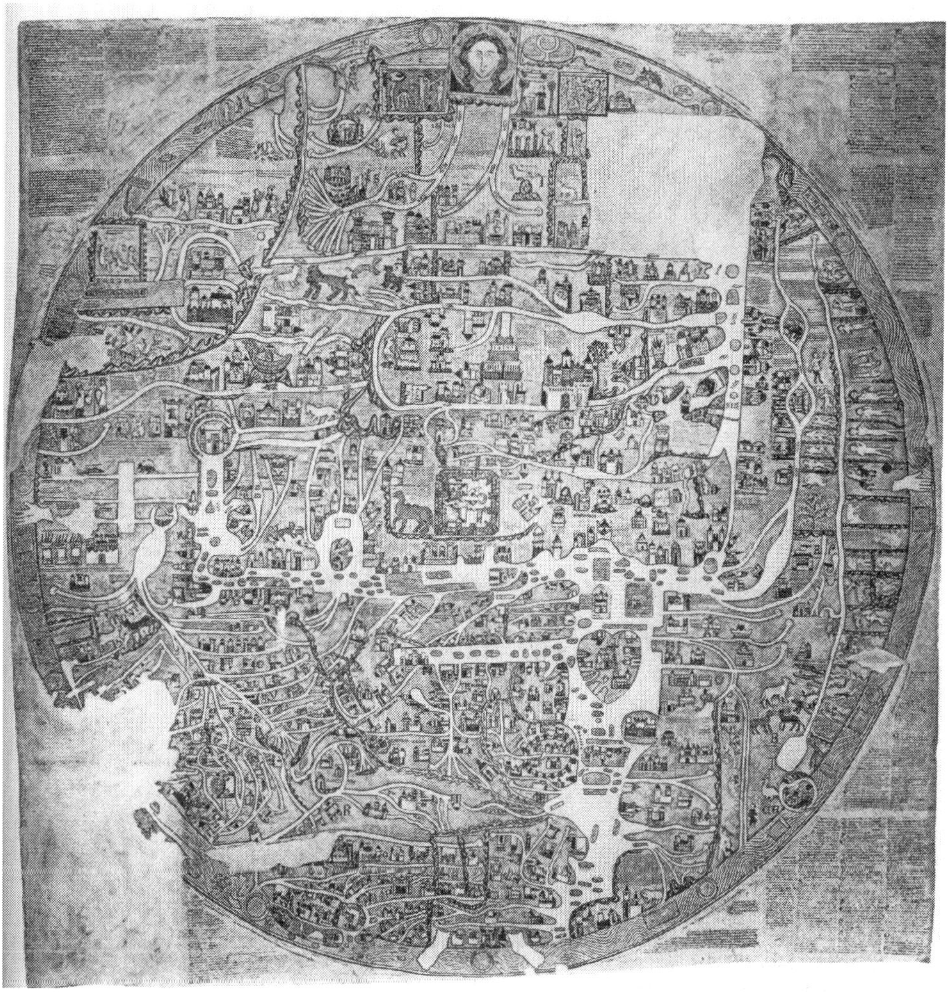

Figure P.6. The Ebsdorf Map, c.1235–39; facsimile version (original destroyed in WW II) from Walter Rosien, *Die Ebstorfer Weltkarte* (Hanover, 1952); see also fig. P.8, below, for details of Christ's head and left hand; reproduced by permission of the Niedersächsisches Institut für Landeskunde und Landesentwicklung an der Universität Göttingen.

Figure P.7. The Creator Measuring the World, from a late thirteenth-century Reims *Bible moralisée*; Vienna, Österreichische Nationalbibliothek, MS Cod. 2254, fol.1; note comparable pin-prick (and concentric compass circles) on the Hereford Map (figs. 2.1 and P.3); reproduced by permission of the Bildarchiv d. ÖNB, Wien.

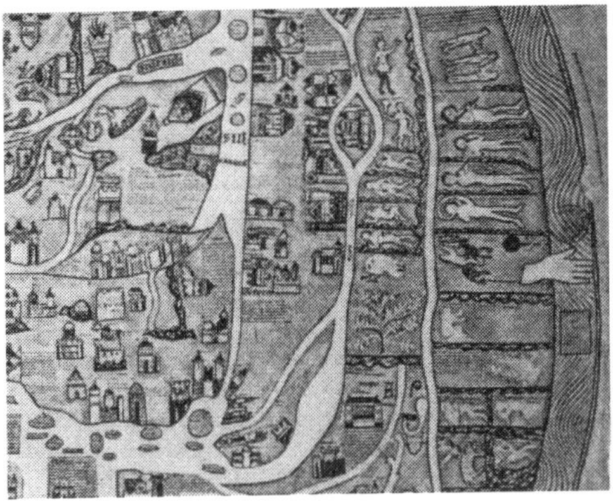

Figures P.8a-b. Christ's Head and Left Hand, details from the Ebsdorf Map (facsimile version), 1235–39; from Walter Rosien, *Die Ebstorfer Weltkarte* (Hanover, 1952); see also fig. P.6 for entire map; reproduced by permission of the Niedersächsisches Institut für Landeskunde und Landesentwicklung an der Universität Göttingen.

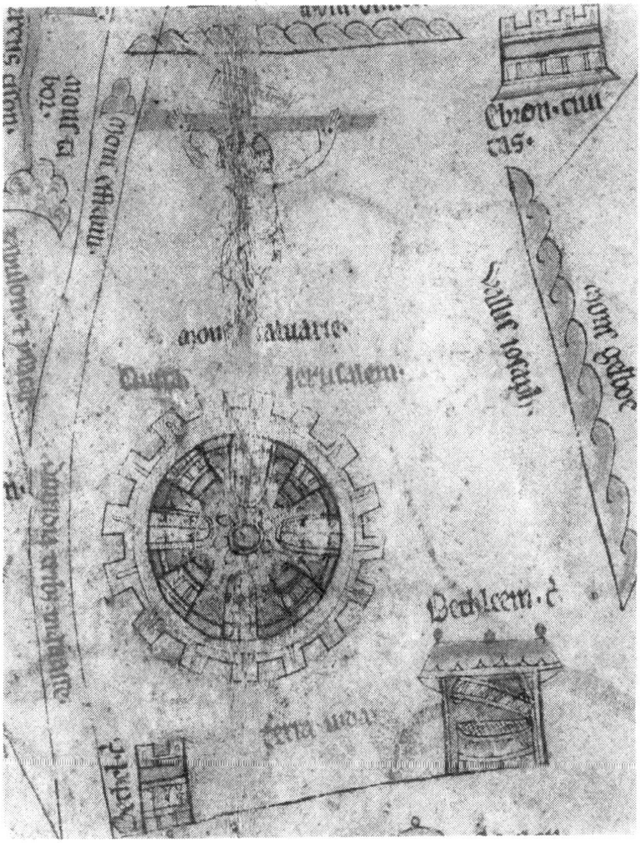

Figure P.9. City of Jerusalem/Crucifixion Scene, detail from the Hereford Cathedral Map; reproduced by permission of the Dean and Chapter of Hereford and the Hereford Mappa Mundi Trust.

THE KING'S TWO MAPS

Chapter One

Painted Chamber *Compilatio*

Henry III and the Cartography of Kingship

THIS CHAPTER PICKS UP WHERE THE PROLOGUE LEFT OFF BY EXAMINING King Henry III's lost Westminster Palace *mappamundi*—commissioned in 1236 as apparent centerpiece, I argue, of an evolving mural program for the Painted Chamber—in the context, first, of contemporary politics: specifically, Henry's attempts to consolidate his power as king against the encroachments of his baronage. Perhaps more importantly, the chapter reads Henry's map—or more precisely, it proposes outlines for a necessarily speculative reading of this ghost of a document—in the context of the visual and discursive conventions (sometimes equally ghostly) that appear to have governed (underlain, constituted, framed, informed) the cartographic practice of the day. Compared with other eras less tolerant of deviation from an accepted standard of cartographic authority, England in the thirteenth century possessed a cartographic culture that was still establishing itself in many respects. Mapping dwelt, that is, in a state of uncertainty, of technical and hermeneutic flux, even if aspects of some generic traditions (as reported in the prologue) were reasonably well-fixed, by the standards of manuscript culture. One result of this situation is that the various formal and conceptual elements that went into making maps (as well as making sense of them) were highly susceptible to ad hoc realignment, to reconfiguration along new and potentially revolutionary epistemological lines. Given such a climate—in which maps remain in the process of developing tangible conventions, and of accruing unto themselves associations, effects, and meanings appropriate to their increasingly settled yet only recently constituted forms—it becomes crucial to ensure that there is critical intercourse with realms beyond the strictly cartographic. In order to determine the structural underpinnings or interpretive valence of any given medieval map-text, that is, we must do more than read maps primarily in the context of one another, as is still often the case whether comparisons range horizontally (across a given era) or vertically (along lines of teleological development). Certainly, such perspectives *are* indispensable to any cultural reading of cartography; the trick must be to strike an effective balance between the practice of comparative cartography and other modes of historicism, geographical and otherwise.

The map-image lying at the heart of this chapter is one that appears to have been available for viewing for a period of just over a quarter of a century—three-quarters of a millennium ago—but was then destroyed or replaced, leaving behind no fully authoritative copy, no detailed textual description, and no direct archaeological trace. Given the unconventionality (not to say inconvenience) of having such a gaping absence at its center, my foray into cartographic analysis in this chapter will have more need than usual to balance close reading of maps (and of contextual documentary information) with creative use of various other orders of evidence. To read an "artifact" so long lost we shall be obliged to ask exactly how a thirteenth-century world map draws the discrete elements of its form and meaning from that wider pool of images and concepts comprising medieval culture.

The risks here are considerable, yet it may be that an interpretive virtue can be fashioned of this methodological necessity. In any case we shall push aggressively outward from consideration of maps alone and in themselves to treat such disparate subjects as early heraldic and seal-making practice; the illumination of medieval Psalters and other books; and finally a number of implicitly geographical written texts, such as a medieval copy of a Late Roman "flowchart" of imperial administrative precedence. Each of these topics will amount to a way-station in our movement toward reconstructing the likely form, features, and display-setting of the king's Westminster *mappamundi*. Yet in our attempt to execute such an act of visual and social archeology, two extant medieval images will take on special importance, on grounds that they constitute potential witnesses to King Henry's lost Painted Chamber mural. These "star witnesses" of mine are the sibling miniature Psalter Maps, whose execution (by different hands but back to back upon a single manuscript leaf) has been localized to the vicinity of London/Westminster, c.1260–65. In other words they were produced at some point just before or after the king's Westminster *mappamundi* was destroyed or painted over. Despite certain problems associated (unavoidably) with their use, these images shall inform (if not direct) major portions of our reconstructive undertaking, insofar as together they provide a working model of Henry's map—if admittedly a hypothetical model, and one complicated further by the Psalter Maps' tiny size, as well as by the seemingly contradictory testimony the two images appear to offer. Still, from a position thus textualized (or visualized), we shall have basis to speculate—on the subjects especially of this postulated production's overall thematic nature and apparent political inflection.

Considering a key shift taking place during the thirteenth century in conceptions of English kingship, the chapter argues, ultimately, that the Painted Chamber world map, by combining geographical data of a Christian sacred as well as of an imperial secular nature, spoke to a newly developing and decidedly composite (or dually based) mode of medieval kingship. The king's Westminster *mappamundi* described a particular kind of royal authority: one that derived its justification at once from the realm of the sacramental and from the realm of the administrative. What is more, taking cues from the layout and content of written codices as well as from painted and illuminated image cycles, Henry's mural map performed this act of political compilation in a strikingly hybrid formal mode, drawing upon constructions of geography which were themselves alternately visual and textual—and then recasting them as simultaneously so.

A WORLD MAP AT WESTMINSTER

How to begin the cultural analysis of a wall painting lost more than seven centuries ago?

Given its destruction (and on top of this the unfortunate disappearance of Matthew Paris's ordinal copy), it would seem that our immediate and most important question must be: What will Henry's Painted Chamber *mappamundi* have looked like, exactly? (Or rather—acknowledging this as an act of critical projection—what might we responsibly *imagine* "the Lord King's map of the world" to have looked like?) Yet never mind, for the moment, the specific features of this Westminster map. A better question to begin with, in its being more precisely answerable, is: What will any given map of the world have looked like—have meant—*at Westminster*?

The precise administrative function and social meaning of Westminster in the thirteenth century appears to have changed somewhat, decade to decade and even year to year, depending especially on fluctuations in the relative power of royal and baronial parties. But one constant, upon which commentators seem to agree, is that the thirteenth century saw an increasing "concentration of ultimate political and governmental power at Westminster." While it would be some time before a London/Westminster grouping came to assert an unassailable cultural and administrative hegemony, and while Rodney Hilton, for one, sees thirteenth-century England as "an inevitably decentralized state," a number of scholars have sounded variations on a theme to the effect that "government and commerce were increasingly centered on London" during the period. With help from the Annalist of Waverly, Maurice Powicke provides anecdotal evidence, in telling how, after having spent Christmas together at Winchester in 1279, King Edward "went off to hunt" while his Chancellor Robert Burnell "returned to London 'as to the fixed place where those who seek writs and pursue their rights might have the appropriate remedy.'" The relocation of the royal treasuries in the course of the century, from Winchester to various sites in London and Westminster, bears further witness to precisely this administrative change.[1]

Expertise of all kinds centered on the king's court, but the royal household in particular, an "essential adjunct to kingship" in Chris Given-Wilson's view, stood at "the political heart of the kingdom." Given-Wilson notes that the king's "affinity" or body of servants and followers is "best envisaged as a series of concentric circles," moving outward from the *domus*, or permanent royal household, to the more expansive *familia regis*, of less fixed membership, and ultimately to a wider grouping of officials and part-time retainers. That the kingdom's political geography centered (most unequivocally) on the person of the king himself was given legal reality in the concept of the "verge," a floating territory of special jurisdiction that extended outward in a radius of twelve miles from the sovereign, wherever he might be.[2]

The royal court, of course, had no single fixed location: king and *domus* were rather "almost permanently itinerant." Yet the king's inevitable removal after a few weeks or months, from one favored residence to a next, ought not to be interpreted as constituting a kind of exile for the monarch from the physical and institutional focal point of the realm, as in the modern sense of a nation-state's capital. Perceived along the lines of the military or juridical circuit, ultimately medieval royal perambulation does not undercut so much as reinforce, arguably, the concept of a political and social center. An official metropolitan presence, that is, whether embodied in royal couriers, commissioners, and tax collectors or by sweep of the king and his court itself, does not make the periphery less peripheral. Rather, such agents of the political center reinforce the notion that executive power extends here, into each regional locale, though its proper home lie elsewhere.[3]

In stressing that it was the realm's "political hub" as well as "the first and most important extension of the king's will," Given-Wilson has characterized the royal household as "both centrifugal and centripetal" in its workings. This spatial model is one we will have occasion to return to later, but for the time being it may be useful to visualize a time-elapsed figuration of the court's perennial radiation between core and periphery. For while the king's court seems to have been importantly and not just incidentally peripatetic, modern accounts and diagrams that reconstruct thirteenth-century royal itineraries describe a decided Westminster center of gravity for this body's orbit. Moreover, although the movements of early Angevin rulers were near frenetic, later English kings peregrinated at a declined rate and across a narrower field, so that by the mid-fourteenth century only seldom did king and household remove themselves more than a day's ride from London, and even in these cases it was "usually for some specific purpose."[4]

As noted in the prologue, a production on the general (portable) order of Edward I's *pannus ad modum mappe mundi*—a tapestry or oversized manuscript, a painted cloth or folding panel—might be the map format most appropriate to the itinerant lifestyle of any medieval English king. Still, a map fixed at Westminster in the mid-thirteenth century (especially one enjoying prominent display) was arguably an icon, on the other hand, which would have been understood as standing at or near the accepted "center" of a coherent social and physical body known as "England"—an extra-feudal entity coming increasingly to be conceived in centralized political terms, if not yet in fully territorial and state bureaucratic ones.

Consideration of the particular role of the Painted Chamber within a larger Westminster complex provides another approach to Henry III's lost map—whatever its own formal details. After the Great and Lesser Halls, the Painted Chamber (fig. 1.1) seems to have been the next most important room in the palace. Much of this importance may derive from its having served, until the reign of Edward II (1307–27), as the royal bedroom. Although it was slept in, the Westminster Great Bed—like the later French *Lit de Justice*—functioned primarily as "a piece of state furniture." Within a decorative, canopied enclosure it comprised not just the monarch's "inner bed" (for sitting and lying on), but an "outer bed" as well, making the whole a kind of "chamber within a chamber." Because access to the king's person meant access to power, the royal bed and larger Painted Chamber thus served, in both social and symbolic terms, as a primary focal point for court activity. They stood, as it were, at the heart of the heart of the realm.[5]

The Painted Chamber had functions beyond its role as the king's bedchamber. From perhaps as early as 1244–59, it was used as one of the meeting places of Parliament, and from the first half of the fourteenth century at least, for state feasting. What is more, it served periodically in the thirteenth century as the site of "major state events" including Alexander of Scotland's homage to Edward I in 1279. Perhaps to lend some coherence to its variety of functions, Henry and Edward saw to it that this room was decorated most carefully, extensively, and formally—and did so with such an ongoing outlay of expense as to make it, in the judgment of Tristam, "the finest of the various richly painted rooms" to be executed in the period. "It had in its day," concurs Pamela Tudor-Craig, "the most important, and no doubt the most influential ensemble of domestic wall-painting in England."[6]

Three major mural-painting programs were undertaken in the Painted Chamber during the thirteenth century. Henry III commissioned the first two and Edward I ordered the third. Begun in the mid-1230s (although conceivably incorporating existing work), the original scheme featured the king's *mappa-mundi* together with a calendar, a mysterious *magna historia*, and a bestiary, to which Henry added the Four Evangelists in 1243 and a Jesse Tree in 1259. In the mid-1260s—after a 1262/3 fire had destroyed or significantly damaged the original paintings—a second, very different mural scheme was undertaken. This full-scale reconception and repainting focused upon St. Edward the Confessor, the pre-conquest English royal saint for whom Henry had developed a deep attachment, and for whom his eldest son had been named. The third scheme (executed under Edward I, c.1292–97) arranged an extensive series of Old Testament illustrations, many of them battle scenes, in narrative layers above the 1260s murals. The attention, expense, and ideological ambition they put into the Painted Chamber's decorations suggest Henry and Edward well understood the importance to political theater of an impressive set or backdrop. Murals like these placed a king in proper context. And the royal bed in particular provided a permanent "architectural focus" for the room—"around which a systematic iconography could be composed."[7]

The business of interpreting the iconography of the Painted Chamber is largely the business of reconstruction. For the latter two schemes, primarily this has meant piecing together the visual evidence provided by nineteenth-century copies of the (then) partially preserved murals with documentary evidence about image content, image placement, and expense. For the earlier scheme, on the other hand, we have no visual evidence whatsoever (owing to the 1262/3 fire and subsequent repainting) so interpretation—necessarily conjectural—means reading lost images according to the evidence provided by their context and by their extant visual relatives.[8] It may also prove useful to read from the 1260s scheme backwards. That is, even if differently conceived and iconographically realized, King Henry's two schemes may have forwarded similar (if differently inflected) basic ideological agendas. What is more, given the chamber's ongoing architectural realities—its floor-shape, entrances, windows, fireplace—it may well be that Henry's two schemes were laid out with a corresponding visual syntax, in the sense that basic relationships between images in a cycle (priority, dependency, opposition) will be implied by location as well as position relative to one another.

The foundational work of Paul Binski in *The Painted Chamber at Westminster Palace* (1986) forms the basis (developed in far greater authority and detail) for much of the account given above. According to Binski, Henry III had a "precise sense of image placement" and "thought in terms of a central repertoire of images" which, for the most part, "remained constant throughout his life." Late in life, however—dating apparently from a 1254 "cultural expedition" to Paris—Henry turned toward a new set of decorative styles and themes, as his wholesale reconception of imagery for the post-fire Painted Chamber helps underline. Yet whether the images Henry chose for his second mural scheme were themselves new departures or instead consistent with earlier interests, in terms of its careful selection of subjects and their precise arrangement this ensemble worked consciously to sketch what is in effect a single "cumulative image"—one which figures royal authority. The 1260s Painted Chamber provides, that is, "a cogent illustration of Henry III's view of his kingship."[9]

St. Edward the Confessor dominates this scheme, as noted above. An elaborate version of the Confessor's Coronation—Binski calls it the chamber's *pièce de résistance*—occupied the wall behind the royal bed (fig. 1.2). Curtains or posts, canopies, and the plane of the bed itself framed the composition, serving to bring it within the ambit of the outer bed (fig. 1.3). In this way a very direct association, between the dominant figure in the image on the wall and the dominant figure on the bed before the wall, must have been achieved. Moreover, a coronation image for the wall behind the canopied royal bed makes much iconographic sense, since "canopies supported on posts were held over the King and Queen during the ceremony of coronation itself."[10]

All told, his choice of iconography throughout the 1260s scheme illustrates, especially in its orientation around the royal bed, Henry's "desire to root his kingship in the person of [his] sainted royal ancestor." The scene of his patron St. Edward's anointing and coronation displayed on the wall behind him suggested that in his own anointing Henry had inherited unequivocal divine support for his authority.[11]

A CRUSADER ATLAS, OR, MAP GENRE OVERLAP: FUNCTION

Binski's reconstructive work dwells most extensively (as is practical and appropriate) on those images from the three Painted Chamber mural cycles for which most evidence exists—i.e., the later ones, those copied and/or described by antiquarians before being destroyed in turn. Unfortunately for the history of cartography, this means that the potentially definitive role of the initial scheme's *mappamundi*—with respect to the meanings incarnated in the room as a whole and over time—remains a subject more suggested, in this magisterial work, than fully elaborated. In other words, Binski declines to analyze *mappaemundi* in relation to the Painted Chamber's apparent function (to judge by subsequent cycles) as a site for the expression of thirteenth-century ideologies of kingship. In the absence of definitive archeological evidence—such as infrared or palimpsest analysis of the central mural depicting St. Edward's Coronation, now obviously impossible—we cannot be fully certain about whether, as I believe, Henry's early-commissioned world map occupied that prime and seemingly most privileged location in the chamber, of the wall behind the royal bed. I take this to be the case not (only) due to wishful thinking, but because no other candidate from among the known 1230s-1260s mural-subjects possesses so strong a claim to likely habitation of this focal position in the cycle. There is the conceivable exception to this of the *magna historia*. However, the term "great history" may (per Binski's suggestion) refer itself to the world map known definitively to have been present and prominent in the chamber, given that *mappaemundi* frequently accompanied or were even described themselves as "histories" (cf. the Hereford Map's reference to itself as an "*estoire*"). Given the subjunctive mood of my inquiry, it might be counterproductive to insist too strenuously upon the point of the world map's central location in the 1230s scheme, at least before all reports are in, all strands of supporting evidence woven together. But whatever spot Henry's map took here at Westminster, or for that matter occupied at Winchester Castle Great Hall, where significant overlap with the themes and subjects found in the Painted Chamber may be observed, what is most important to recognize is that *mappaemundi* had apparent policy implications in the thirteenth century. Specifically, they seem to have played a key role in the period's negotiation of the nature of English kingship—particularly in terms of the spatial (geo-religious and geo-political) responsibilities of the office.[12]

Political histories tend to read the English thirteenth century primarily in terms of a conflict between the two poles of "royal absolutism" and "baronial rebellion." Also, many have noted the "unpopularity of overseas expeditions" during the reign of Henry III. Conflict between royal and baronial partisans manifested itself most bitterly in precisely this context, of the king's overambitious and expensive continental adventures. Despite the often equivocal commentary which they accompany, contemporary maps register implicit support for royal military projects abroad (this in contrast with their general unpopularity) and for the idea of a conquest-oriented politics generally, by outlining, as they arise, the opportunities for territorial acquisition available to an ambitious royal house well-placed in the aristocratic network of feudal Europe. Thirteenth-century maps, that is, illustrate the spatial implications of royalist (as opposed to baronial) interpretations of English kingship. As an example, let us consider "the Lord King's map of the world" in a slightly different context: not on display in the Westminster Painted Chamber, but instead copied into the lost ordinal of Matthew Paris—chronicler, cartographer, and monk of St. Albans.[13]

Matthew Paris connects closely with Henry's court on a number of grounds. For example, his visual work overlaps considerably, in terms both thematic and stylistic, with that executed in the mid-century at Westminster and other royal decorative sites. Although Paris died in 1259, it has been suggested that he may originally have designed the St. Edward scenes executed for Henry's 1260s Painted Chamber, including the Confessor's Coronation. Of more immediate interest, Paris drew a large number of maps, in a variety of genres: maps of the world, maps of regions (e.g., Palestine and Britain), and plans or detailed views of cities (e.g., London, Rome, Jerusalem, and the crusading capital of Acre), as well as four versions of a remarkable itinerary map from London to Apulia in southeast Italy. He is thought by some to have inspired the production of a local map (a plan for installation of an underground water pipe at Waltham Abbey); and it is even conceivable (so Tristam suggests) that Paris himself may have designed—not just "copied" into his ordinal—Henry's Westminster *mappamundi* itself. This seems unlikely (given the readings Vaughan and Lewis have given of this passage) but even so, Matthew's overall level and variety of cartographic production astonishes—the more so for a period so scarce of maps, and for one in which "to draw any map at all was a bold conceptual initiative."[14]

Cartography as an intellectual and artisanal activity does not separate readily, in these years, from practices related to it, such as wall and panel painting, mosaic and tapestry design, manuscript illumination and copying, chronicle or treatise writing, and so forth. According to some, maps should probably not even be seen "as constituting a single class of object," as distinct from a textual diagram tradition on the one hand and from painting and illumination on the other. Furthermore, the different cartographic systems presented by different medieval map genres have been typically understood to have "existed in largely separate traditions" and moreover to have had each their own specific, single function. One critic, for example, has observed that "most medieval *mappaemundi* share no obvious formal or functional similarities with other maps of the period." Another goes further: "There was no such thing as a general map designed to be put to a wide variety of uses. Any one map was drawn with one particular purpose, even one particular occasion, in mind."[15]

Considerable evidence, especially that related to a few outstanding medieval cartographers, suggests a need to revisit these conclusions concerning non-overlap between purportedly discrete map-genres and their apparently specific uses. For example, that one of just two "local maps" extant from all of thirteenth-century England—the above-mentioned plan for installation of a water-pipe—not only connects indirectly with Matthew Paris but comes from the same abbey, Waltham in Essex, which housed a famous *mappamundi*, provides evidence that a general and flexible concept of the map existed, straddling cartographic traditions usually seen as both formally and functionally divergent.[16]

In the early fourteenth century Pietro Vesconte, a Genoan in Venice, produced numerous maps in, like Matthew Paris, a variety of genres. But Vesconte also drew portolan charts, nautical maps of everyday use and increasing technical precision whose underlying principle of empiricism, along with the rediscovery of Ptolemy's *Geographia* in the West, would help revolutionize cartography in early modern Europe. In c.1320, Marino Sanudo compiled several of Vesconte's productions—maps of the world and of the Holy Land, and plans of Jerusalem and Acre, all inflected with portolan-chartmaking techniques—into his *Liber secretorum fidelium crucis* ("Book of the Mysteries of the Defenders of the Cross"), a tract written "as a means of arousing interest in a crusade." The compilation was presented to Pope John XXII in 1321, in neither the first nor the last attempt by Western scholars to impress upon the pope that reconquest of Jerusalem depended upon a systematic understanding of geography.[17]

Mappaemundi had been associated with Christian evangelism and militarism long before the thirteenth and fourteenth centuries. The Hereford Map, for example, derives ultimately from a prototype found in manuscripts of Paulus Orosius's fourth-century *Historiarum adversum paganos* or "History against the Pagans." The Ebsdorf Map and its Isidoran prototypes and analogs also contain what has been seen as crusade-propaganda, and it is significant for concepts of a universal church that Christ's left hand draws even the monstrous races into the community of his body (see fig. P.8, above). *Mappaemundi* associated with Beatus's *Commentary on the Apocalypse* seem to have had an especially explicit evangelical or apostolic function, as indicated in some cases by the presence of the Four Evangelists in the map's four corners, in a kind of reprise to the representation on some T-O maps of the dispersal of Noah's three sons to the three known continents: Shem to Asia, Ham to Africa, and Japheth to Europe (see fig. P.5, above). Given these longstanding evangelical associations, and the genre's increasing tendency to place formal emphasis on the Holy City as the physical and spiritual center of the world, the inability of Western Christendom to hold Jerusalem made the medieval T-O map "a matter of considerable embarrassment" to devout Christians overall, but especially to likely (or lapsed) crusaders.[18]

Marino Sanudo's use of multiple maps by Pietro Vesconte, and not just his *mappaemundi*, suggests that despite their formal differences medieval maps of the various genres (world, region, city, single itinerary, even portolan chart) were in fact seen by contemporaries to share a degree of basic functional complementarity. Each contributed to what might be called a crusader atlas.

Like Vesconte after him, Matthew Paris copied, or drafted himself, maps of all the types that might be useful for the symbolic exhortation to crusade, or its practical planning out. His maps of the world, his itinerary maps to and through the Holy Land, and his plans and views of cities including Jerusalem, Damietta, Damascus, and Acre all participate, in much the same fashion as Vesconte's do for Sanudo, in "the great expedition" undertaken by Western Christendom and especially its princes, including from the English royal house Prince Edward (1270–73) and before him his uncle, Richard of Cornwall (1239–41).[19] Paris's maps, in other words, although a number have been lost and those extant are scattered across several manuscripts, prove coherent and complementary overall in that they provide technical and ideological support for both crusade *per se* and for certain overseas adventures of the royal house connected with it.

For example, Matthew's Itinerary from London to Apulia appears to be connected with Pope Innocent IV's offer, in 1252–53, of the crown of Sicily (including Apulia) to King Henry's brother Richard of Cornwall. After Richard colorfully refused ("You might as well say I will sell you the moon; go up and take it") Henry arranged for his second son Edmund to receive the crown, in exchange for which he was to send troops and money to Italy. In 1255 Pope Alexander IV commuted the vow Henry had taken in 1250 to go on crusade into an obligation to further the church's war against its imperial enemies (the Hohenstaufen) in Italy—the advantage being that the funds collected from the English clergy since 1250, in support of Henry's vowed crusade, could in this way be used in support of his son's claim to Sicily, as this too was now designated an ecclesiastical enterprise. By all accounts Henry overstretched himself badly. Edmund had the title but not the territory; to win it would require a staggering outlay of both funds and personnel on the part of an English baronage who not surprisingly "had little interest in the adventure."[20]

Matthew's *Chronica Majora* describes the papal offer of the crown to Richard, and in addition to glossing sites from the earl's crusade (e.g., his landing at Trapani in 1241) one copy of the itinerary contains an inscription which outlines Innocent's proposal, in seven lines of text running across and around the legend *Poille* or "Apulia" (see fig. 1.4, absolute top and upper right center). What the itinerary map provided was the route to English conquest of Apulia, in the form of a series of vertical strips setting out the staging points in such a journey. Essentially this meant a list of the names of towns, with distances in days and a line marked *Jurnee* between them, together with architectural sketches and bits of other relevant information such as rivers and bridges (see figs. 1.5 and 2.5, below; at some points there appear alternate routes).

The outstanding formal characteristic of the itinerary map is its clear linear focus; its spatial argument admits of no deflection or reversal but drives the eye and imagination insistently forward. As Michel de Certeau has said of medieval itineraries, "each of these maps is a memorandum prescribing actions." In its methodical sweep from London through Rochester, Canterbury, Dover, and so on, day by day, to Rome and finally Apulia, Matthew's itinerary implies that practical realization of the pope's grant of investiture was all but assured, English success a matter of course. In short, whether on behalf of Henry's brother or his second son, English royal passage to conquest of the Kingdom of Sicily was, literally and figuratively, mapped out.[21]

At one time scholars regarded this work as an itinerary from London all the way to Jerusalem. However, the Map of Palestine which follows the London-Apulia Itinerary in three of four versions, and which structures this territory along a route through Acre (the last crusader town held) to a centrally placed Jerusalem, probably constitutes a separate work. (In fig. 1.6 a camel points the way, while in another version distances between towns along the coast are marked.) The two maps blend together, however, not just in terms of their adjacent geography and sequential manuscript placement. For to align the passage of an English king-elect to the conquest of Sicily with the greater expedition of a unified Western Christendom's passage in arms to Jerusalem, is to underline the inflection of crusade which the pope himself had lent the "Sicilian Business" (*negotium regni siciliae*) by substituting it for Henry's promised crusade. To pursue the crown of Sicily was, in other words, not simply an expensive secular or feudal ambition of the English royal house. As argued by the manuscript alignment of Matthew's London-Apulia Itinerary with his Map of Palestine (organized itself along an itinerary to Jerusalem), Henry's papally sponsored but domestically unpopular "Sicilian Business" partook of the larger, sacred duty to re-win the Holy Land which, it is said, "reverberated reproachfully" in the religious conscience of every devout Christian of the period.[22]

We noted in the prologue that maps and secular rulers link closely. In fact, cartography links kings directly to crusade. Commentators tend to imply that the perceived duty of crusade fell evenly upon medieval Christians—in that "the business [*negotium*] of the Holy Land was a political and economic function of society" as a whole. However, as with cartography, crusade associates with royalty much more closely than it does with other groups of elites, such as the baronial or mercantile. Simon de Montfort, for example, was "unusual among the magnates in being associated with the evangelical Christian movement of his time." In contrast, Queen Eleanor's "Antioch Chamber" at Westminster Palace, decorated with scenes from the First Crusade, as well as King Henry's repeated commission elsewhere of paintings of his uncle Richard Lionheart in combat with Saladin, suggests that for the royal household "a visual reminder of the centrality of the crusade and the glory of the crusader" was never far away. Henry commissioned his Westminster map of the world in the same year (1236) that saw a crusade preached in the west. Richard of Cornwall "took the cross," as Henry would himself in 1250—a year in which Matthew Paris had much to report of crusading activity in his *Chronica Majora* and in which (not coincidentally it may be) he chose to sketch at least one surviving world map, while referring to his careful reproduction of another. Where the world map copied by Matthew Paris from the Painted Chamber mural cycle may have expressed Henry's eventual intention to go on crusade, meanwhile serving in part as a substitute for it, the London to Apulia Itinerary described the holy expedition in Italy that the king's crusade vow would develop into by the mid-1250s.[23]

Medieval *mappaemundi* in particular, but other medieval maps as well, detail the spatial duties and proper geopolitical aspirations of the Christian monarch, specifically his obligation as Vicar of Christ to concern himself and his kingdom's resources with the pressing business of the recovery of Jerusalem and related projects. In support of both these enterprises maps trade on an association between the body of the king and the body of God or Christ, which is also to say, of the created world, in order to describe Christian kingship as sacramental, rooted in the all-encompassing authority of the holy. That is, because a king's authority derives directly from divine authority, to go on crusade reinvests directly in, and in fact marshals, a king's own power, even as it fulfills an essentially feudal obligation to his spiritual overlord.

That *mappaemundi* describe English kingship in this way may explain why Henry III commissioned a map of the world for his Westminster Painted Chamber in the same year that his brother Richard of Cornwall took the cross (1236), and why he commissioned another for Winchester's Great Hall in the year Richard was finalizing plans for his departure (1239). It may explain why Richard or his son Edmund, next Earl of Cornwall, commissioned production of the Duchy of Cornwall map (c. 1260–83), and why Edward I, "flower of chivalry" and the "last great crusader of his age," would have been particularly appropriate as patron for or recipient of a *pannus depictus ad modum mappe mundi*. This reading also suggests why Henry may have chosen not to repaint a world map in the Painted Chamber in the 1260s. The Sicilian Business had been a domestic disaster for him, and led directly to his humbling at the hands of the barons in 1258, under the Provisions of Oxford, and eventually to the Barons' Wars of the early and mid-1260s. Intriguingly, evidence suggests that the civil war "almost brought about the destruction of the Painted Chamber"—who is to say how consciously, with regard to its iconography?—in that the king's opponents "broke the doors and windows and scarcely withheld their hands from the burning of the whole palace." By the time of the Chamber's repainting, it was clear that Henry would never undertake the crusade he had vowed in 1250. *Mappaemundi* outline the geopolitical responsibilities of a Christian monarch, and suggest his proper aspirations. But "the Lord King's map of the world" described an itinerary Henry failed to take.[24]

KINGSHIP, PART ONE: SACRAMENTAL AND LITURGICAL

An appreciation for the "intimate connection of politics and religion" dominates early medieval political thought. In *The King's Two Bodies: A Study in Medieval Political Theology* (1957), Ernst Kantorowicz argues that the basic terms according to which kingship was conceived changed during the thirteenth century, for England in particular. Whereas later, medieval kingship's dominant modes were essentially administrative, primarily legal and "fiscal," Kantorowicz characterizes early medieval kingship as most importantly "sacramental" and "liturgical," terms that have to some extent informed this discussion already. As Kantorowicz describes it, this earlier model was a "definitely Christocentric doctrine of kingship," in which the "divine prototype" and his "visible vicar" were thought to reflect each other, and thus to "display great similarity."[25]

Coronation practices play an important part in Kantorowicz's readings of medieval kingship. In the early medieval model, the anointing of the monarch at his coronation defines kingship as nothing less than the "effluence of a sacramental and liturgical action performed at the altar." In other words, a king "becomes deified by consecration" and only through this ceremony did he take on, by grace, the power of God and Christ.[26] Sacred authority underwrote, through coronation, the exercise of secular power.

Much material links Henry III to liturgical kingship. For example, early in his maturity, Henry responded to a baronial challenge to his authority in the localities by mandating an increase in the use of the *laudus regiae*, an ancient liturgy of praise to the ruler. This increase reinforced the notion that the king's exercise of power, because rooted in liturgical practice, had divine sanction, and helped justify sweeping administrative changes in the direction of centralization. Similarly, a practice whereby the kings of England and France sought to heal by a laying-on of hands those suffering from scrofula, commonly known as *le mal du roy* or "the King's Evil," also describes Henry's kingship as essentially sacramental, in its concern to "underline the spiritual qualities and functions of kingly office." The practice appears to have developed in the latter years of Henry III and Louis IX, and may have passed from one to the other during Henry's visit to Louis's court in 1254.[27]

Kantorowicz's observations on liturgical or sacramental kingship help identify the precise links which Henry's map of the world likely suggested between divine and royal authority, the melding we may speculate that it argued between the nature and power of Christ and that of his royal vicar on earth. As acknowledged above, it is impossible to say with full confidence and authority what exactly this map looked like (and so we deferred the question). Yet figure 1.7, which reproduces the full-page manuscript illumination known as the Psalter Map, and figure 1.8, which reproduces the lesser-known illustration occupying its verso, may partly answer this. One or both of these miniature maps may bear witness to Henry's lost Westminster prototype, as noted above (both date to c.1260–65). Or recto and verso may together, reconstituted as it were, comprise a single copy of the mural between them, in two-part composite. Whether these copies are direct or indirect; how other visual sources may have been involved; and how fully these images reproduce a much larger original's features, cannot be said for certain. But there is at least precedent, in the work of Matthew Paris, that establishes the practice of producing "reduced copies" of monumental *mappaemundi* in these years.[28]

WORLD, BODY, MAP

Much in the way of the Ebsdorf Map, the body of Christ gives structure to the Psalter Map (fig. 1.7) and indeed merges with its image of creation. Christ's torso grows directly out of, as it were, the roughly tripartite disc of the earth. In the Psalter Map's verso illustration (fig. 1.8), moreover, God or Christ (it is unclear which) holds a second T-O map before him, his fingers curling along its edges and his ankles and feet emerging into the space below. Ebsdorf's incorporation of Christ's body into the very texture and material topography of the world goes further than do either of these images, and the Ebsdorf Map distinguishes itself also in strongly emphasizing the crucifixion (figs. P.6 and P.8, above). Still, the recto and verso Psalter Maps directly invoke the *mappamundi* form's most arresting arguments in this period—not only that in his incarnation Christ advances the movement of sacred history begun in the Creation, continued in the Fall, and to be completed in the Resurrection and Last Judgment (events likely included in the Painted Chamber's *magna historia*?) but also, and perhaps more centrally, that Christ's body itself comprises the basic substance of creation, in effect doubling as the body of the world. Especially compelling in this context is the (palm-sized) Psalter Maps' precise Eucharistic circularity. In the case of the recto, no fewer than five concentric circles share Jerusalem as their center point: two give body to the Holy City, a third marks the outer edge of the three known continents, a fourth draws the band of an encircling ocean, and a fifth provides a border, into which the winds and their names are set, between the map-image and the rest of the manuscript page.[29]

At nine centimeters across the recto Psalter Map has little enough space for representation, yet a few additional features stand out and point to likely avenues for interpretation. Psalter's collection of monstrous races in Africa, for example, suggests that an evangelical argument may have been especially important to Henry's larger mural map, a point particularly resonant in that a bestiary had been commissioned to accompany it originally, and given, what is more, the addition of the Four Evangelists to the Painted Chamber (one upon each wall) in 1243. In another vein, the miniature map's twelve winds, in that they are drawn at precise intervals along the world's border, may indicate a degree of interest in abstract direction (after the manner of the later portolan chart "wind rose" and of Ebsdorf's cruciform cardinal poles of head, hands, and feet) or in other aspects of scientific geography, such as climatology, meteorology, astronomy, or astrology.[30]

Yet the recto and verso Psalter Maps' most salient features may be those inhabiting the margins outside their map-images proper. For example, where Christ hangs or lies crucified on the Ebsdorf Map, pinned beneath the encyclopedic weight and diversity of creation (and effectively bound by its material limits), on the recto Psalter Map a comparatively liberated Christ—the Christ of the Resurrection—emerges into majesty above the world, or as Michael Camille has put it, he is "outside time and space." Triumphant, he holds his arms up and away, the right in blessing and the left bearing an orb inscribed with a "T," or cross, of the sort commonly held by medieval Christian kings and emperors (fig. 1.9), and before them, Roman emperors and consuls. Although three-dimensional, in effect this is a species of T-O map itself, and symbolizes its holder's sway over (in the case of the Christian Middle Ages) a divinely created earth, one measured and encompassed by Christ's incarnate presence. In the case of Rome, the ruler's orb represents a world geographically known and administered, one scientifically measurable as well as militarily conquerable. That the King of England was argued to reflect the image of the King of Heaven (after the model of the Holy Roman Emperor) may be observed in a number of thirteenth-century compositions which include this detail, among them Henry III's Great Seals of 1218 and 1259 (figs. 1.15 and 1.16, below) and Richard of Cornwall's 1257 seal as King of the Romans (fig. 1.10).[31] What is more, in an arrangement directly reminiscent of English coronation, where censing clergy accompany the king throughout, two angels flank the Psalter Map's purple-robed monarch, and swing censers before a night-sky or tapestry of white stars set in blue, each in a cluster of three.

On the verso Psalter Map, Christ again does not hang crucified or lie pinned by a fallen world, but rather stands erect and active, his hands grasping the disc of the earth in a gesture of physical reclamation (or if God the Father, presentation), and his feet securely on the heads of two wyverns or dragons. These serpents also appear below the disc of the world on the recto Psalter Map, as if to provide foundation for it, although here no actual feet emerge to tread them down. More importantly, wyverns or dragons regularly appear elsewhere under the feet of kings in thirteenth-century wall painting and manuscript illumination. In a curious further twist, the tails of the wyverns beneath the Psalter Map curl and split, as they rise, into sets of what appear to be white fleurs-de-lis, scattered like the stars above and behind the world in a field of royal blue. Medieval wall-painting authority E. W. Tristam regards the wyvern, overall, as "symbolic of the evil spirit, especially of the devil as tempter." But a link to mapmaking traditions generally and some precedent for the wyvern's specific appearance here might be sought either in later (tenth- to thirteenth-century) manuscripts of Beatus of Liebana's eighth-century *Commentary on the Apocalypse of St. John*, where both world map and dragon figure centrally (though not in this proximity), or in the encyclopedist and cosmographer Honorius of Autun, manuscripts of whose twelfth-century *Imago Mundi* include *mappaemundi* as well as a sermon on Psalms 91:13: "And the dragon shalt thou trample under foot." The combination of wyverns and fleurs-de-lis may, as Tristam suggests regarding wall-painting sites elsewhere, have been intended to contrast the state of the Fall with that of the Redemption, since the fleur-de-lis was "traditionally associated with the Virgin" as the operative "human instrument of the Incarnation." By thus emphasizing the Incarnation—through visual reference to the Eucharist, through Virgin-associated fleurs-de-lis, and in Christ's purple robe, evocative of Pilate's *Ecce homo* ('Behold *the man*')—Psalter further stresses that in a basic sense a medieval T-O map connotes bodilyness; cartography implies corporeality. And here, given additionally the unmistakable heraldic association already in this period of the fleur-de-lis with the arms of the French crown and emergent state (and to a lesser extent, the English as well), it does so not just generally but in a specifically royal context.[32]

In short, what most impresses about the recto and verso Psalter Maps is their visual evocation of an earthly king who never actually appears on the manuscript page, but whose presence before and resemblance to a divine prototype—whose "Christomimesis," Kantorowicz would say—is implied through kingship's iconographic shorthand, the royal inflections so to speak, with which the figure of the divine is presented. Although the king himself is absent, in both overall visual scheme and in their choice of details, these sibling Psalter illustrations yet describe him as "the image of the living Christ."[33]

If it appeared, as I believe is most likely, in the space immediately behind the royal bed, King Henry's lost mural map probably measured something comparable to the 1.7 x 3.2 meter Coronation scene which replaced it.[34] Yet as indicated even by a pair of "reduced copies" only 1/20 the size of the original, long before St. Edward argued from the wall behind the royal bed that a king became deified in consecration and that he took on by his anointment a share in the power of God and Christ, Henry made certain to describe his kingship as rooted bodily in the sacred authority of the Incarnation. In effect, our two Psalter Maps propose (should we choose to accept their testimony) that the body divine, the body terrestrial, and the body royal came to coalesce, in Westminster's mid-thirteenth-century Painted Chamber.

THE HERALDIC MAP

To the extent that his commission of world maps in the mid-1230s may have invoked the glory of the crusader and stirred thoughts of knightly ambition generally (but especially in a mixed sacred and secular context), we might say that King Henry, still relatively new to his majority, was here "engaged in the process of fashioning a chivalric identity." Increasingly central to such an identity in the thirteenth century were heraldic arms. For while in many details heraldic conventions had not yet become fully established—in the use of "supporters" for example—authorities agree that the "foundations of a system" were in place by the mid-thirteenth century, and that arms had come already to be "looked upon as characterizing the person." Indeed, "far more than simply a means of identification for the tournament or battlefield," armorial bearings were "a statement both of identity understood genealogically and of social privilege justified by heredity." To be sure, the king occupied a position in the chivalric (and hence heraldic) economy which was defined by an inherent tension: he was at once the "locus of authority" in an increasingly centralized emergent state, as well as a chivalric agent himself, and thus both in conflict with and highly invested in a culture which, feudal and chivalric, "by and large represented a countercurrent" to the movement toward state "so evident in the later Middle Ages." This paradox aside, what is chiefly important for our discussion here is that heraldry, like chivalry, "entailed a form of selfhood insistently, even exclusively, public."[35]

Ann Payne has observed that "from the middle years of the thirteenth century heraldry played an increasingly important part in all branches of artistic activity." We have seen that the Painted Chamber served, from the 1260s at least, as a site for the expression of Henry's basic "view of his kingship." What we have not yet stressed is that, as Payne notes in passing, heraldry was "an important element" in the Painted Chamber's "overall design." This point is easily demonstrated in the case of the 1260s scheme, which included (as we know from the drawings of nineteenth-century antiquaries) "at least one prominently displayed Shield of England," as well as the arms of Henry's brother Richard of Cornwall, emperor-elect and King of the Romans (i.e., Germany) from 1257 (fig. 1.11). It should be noted of these murals, however, that in each case the narrative triumph of virtue over vice dominates; that is to say, in neither case do these armorial devices, in and of themselves, serve as even their individual image's focal point—let alone do they serve as focal points in the larger visual program comprised by the room as a whole.[36]

The presence and relative importance of heraldic elements in the pre-fire mural scheme is much less easily established. Nonetheless it would appear that, if in somewhat less closely regulated and technically systematized terms than in the 1260s scheme, aspects of the basic iconographic logic, or layout, of armorial bearings—in short, the visual language of heraldry—may indeed have played a central role in the initial Painted Chamber mural scheme. And rather than allow to remain bracketed the paradox mentioned above, concerning the king's split political identity—chivalric, heroic, decentralized, and feudal versus executive, administrative, centralized, and fiscal—we might say also that the language of armorial bearings itself provided (at least in the relatively loose grammar of its forms as they were used here) a degree of visual resolution to this basic conflict.

In short, the introduction of an armorial context to our discussion of the Psalter Map recto and verso promises noteworthy interpretive returns. For to "blazon" these compositions (that is, to describe them in heraldic terms) takes our enquiry into the social and political function of the initial Painted Chamber mural scheme one crucial step further. Heraldry provides a mode for reading the

King's commissioned Westminster Palace *mappamundi* as effectively an emblem, even a "characterization," of Henry's public "selfhood," his political identity. This point stands not in opposition to, but rather in useful addition to, its function as an argument for his piety.

It may be no accident that manuscript maps like Psalter and the so-called "Henry of Mainz" or Sawley Map before it (fig. 1.12) bear a marked resemblance in overall layout to the more elaborate medieval coats of arms, particularly later examples and those of the royalty and higher nobility, as exemplified in figure 1.13. According to the technical vocabulary of heraldry, "supporters" are "figures of living creatures placed at the side or sides of an armorial shield, and appearing to support it," while "tenants" denote "human figures or angels similarly used." A heraldic shield may have a number of "lesser charges" (devices or components), but the emblem is defined primarily by the tincture of its "field" and by its "main charge" or primary feature, which is often derived from a pun on the bearer's name (e.g., the French crown's *Clovis, Lovis, Loys, Louis* → *fleur de Loys*, fleur-de-lis). In some cases, several arms or devices might be combined (or "marshaled") in one field, "to indicate marriage alliances, territorial claims, or insignia of office"—a point which (given that Henry III of England and Louis IX of France were brothers-in-law and had numerous other family connections) may help explain the presence of fleurs-de-lis in the Psalter Map's lower spandrels and their proliferation at Westminster in the mid-century generally.[37]

A number of points further reinforce heraldry's particular relevance to the analysis of a world map behind the royal bed in Henry's Painted Chamber. The 1259 addition of a Tree of Jesse, for example—such trees depict Christ's ancestral line through the prototypical King David down to a sleeping Jesse—makes clear the Chamber's explicit concern for the figuring of royal identity genealogically. And though admittedly much later, engraved images of the French *Lit de Justice* and *Lit d'Honneur* illustrate effectively (both were in employ from late medieval times, and had English analogs) that the western monarch's coat of arms came in time to occupy that very spot behind a royal bed into which Henry had placed first his map (I maintain) and later St. Edward's Coronation (fig. 1.14). Somewhat more synchronically, the evidence of thirteenth-century English royal seals also invites application of a heraldic mode of interpretation to the two Psalter witnesses of a Painted Chamber world map. For where Henry's 1218 Great Seal (fig. 1.15) pictured him (somewhat minimalistically) enthroned, feet upon lion and dragon and with sword and orb in hand, his 1259 version (fig. 1.16) not only adds another pair of supporters, leopards erect beside the seated king, but also gives much increased prominence to a structure which, complete with fleurs-de-lis emerging from orbs at its four posts, has usually been identified as Henry's 1245 throne but which also recalls the throne-like royal bed.[38]

Seals, in fact, tie in closely with social and technical developments in medieval heraldic practice. On one hand, because "the representation of the possessor of the seal on the seal" is a convention "older than heraldry," it is agreed that the seal's operative function (as "a mark of authority, identity, and ownership") must have informed heraldry profoundly in its formative stages. On the other hand, heraldry's developing visual language (Hasler describes it as essentially pictographic) seems for its part, reflexively, to have transformed sealmaking: "As the personal, authenticating marks of their owners, seals lent themselves to the introduction of armorial devices which were well-fitted to express a form of signature." In any case a sharp increase in the popularity of seals, not just among the higher nobility and gentry but within the mercantile and ecclesiastic classes as well, helped ensure the development, between the late twelfth and the late fourteenth century, of what may be described as widespread heraldic literacy. For this reason—and "in view of the concept that arms or the armorial beast directly represented the arminger" (i.e., their bearer)—we should very much anticipate the discovery of "near-heraldic beings" on seals and other images relating to individual or collective identity.[39]

From the time of their first recorded appearance on Richard I's Great Seal of 1195, the Royal Arms of England have remained, with occasional additions and minor modifications, fixed and consistent: three leopards in a field of red, or in the language of blazon, *gules, three lions passant guardant.* Simultaneously, nonetheless, there have flourished also "an immense range" of royal badges, iconographically both related and unrelated to the official arms. The terminology threatens to overwhelm, but technical distinction here will prove important: a badge is a "device or mark of recognition" not normally part of the achievement or coat of arms proper; as well as being "a very personal symbol," a badge could also be a communal one, and in fact "need not have its basis in heraldry at all."[40]

Badges did not reach their "hey-day" until the later fourteenth century and following (the "age of livery and maintenance," sometimes known as bastard feudalism) but the same beasts which would later figure prominently on retaining badges (e.g., lion, griffin, dragon, antelope) began to appear somewhat earlier as supporters of the heraldic shield. Supporters would develop over time into an "integral part of the coat of arms, to which certain definite principles" could be applied: typically, for example, only high-ranking nobility enjoyed the right to reserve particular supporters. Before such firm conventions could be established, however, "a long and fairly obscure process of development was necessary"—one which apparently began in the thirteenth century but which remains mostly undocumented, even in the written and painted armorials (encyclopedic rolls of arms), one of the first of which appears (appropriately enough, given his various already-cited connections to Painted Chamber iconography) in the work of Matthew Paris (fig. 1.17). Some commentators point to the roof of Westminster Great Hall, where carved angels carry the arms of Richard II (1376–99), as "an early and important influence on the practice of adopting supporters." However, a better explanation for the emergence of supporters may be found, once again, "in the seal engraver's art; in the decorative devices, *particularly dragons or wyvern-like creatures*, introduced to fill the space between the shield and the circumference of the seal" (emph. mine). The example of figure 1.18, an early fourteenth-century seal bag embroidered with the Royal Arms on one side, and those of the City of London on the other—both supported by dragons—illustrates the basic practice.[41]

The early use of supporters, then, was on one hand comparatively unregulated, and on the other clearly paralleled by the frequent appearance of beasts, human figures, and/or angels on images which were generically related to armorial bearings, but which were not, technically, part of the heraldic achievement *per se*. What this means, for the period before the Black Death and the subsequent age of retaining, is that in many ways the "near-heraldic" details displayed on badges, seals, and other images (especially of identity) will have enjoyed somewhat greater flexibility, and thus more symbolic range, than they would be allowed a generation or two later, when "heraldic art reached its maturity" and strict regulation (by the newly prominent class of professional heralds) and codification (in rolls of arms and heraldic treatises) meant that there was less tolerance for representational ambiguity. Instead, the prevailing aesthetic of heraldry (that which was available for exportable use in a wider artistic realm) depended more and more on a clearly enforced (and at times hotly contested, as in the Scropes/Grosvenor controversy of the late 1380s) one-to-one relationship between the unique armorial signifier and an individual chivalric signified. By the seventeenth century heraldry could be called, on account of its hyper-regulation, "a study which loads the memory without improving the understanding"; but in the thirteenth century it was still very much "a living art," one still developing in terms of its social meanings and its visual conventions.[42]

That Henry III was "apparently personally responsible" for the key alterations made to his second Great Seal, in 1259, provides one relevant example of the flexibility still common in thirteenth-century heraldic practice. The king's decision to add to his seal a new set of supporters, leopards erect as noted above, and to substitute a second pair of leopards (*passant*) for what in 1218 had been one lion and one dragon underfoot, doubtless figured as a key step in the march to the eventual establishment (at least by Stuart times) of leopards as "Royal peculiars"—that is, as supports reserved to the achievement of England's Royal House. Of somewhat more importance to this discussion may be that such leopards were likely meant to duplicate those made for Henry's new throne in 1245, themselves designed to recall "the two lions standing beside the arm rests" of the Throne of Solomon in 1 Kings X:19. Certainly, in this light, the two Soldier-Guardians of Solomon's Bed which would come to flank the central Coronation mural in the 1260s Painted Chamber must resonate, at some level, heraldically (see fig. 1.19 and fig. 1.3, above, far left). But it may be, what is more, that such figures bear witness to these bedside spaces' previous occupation by, say, angel tenants (as on the Psalter Maps) or other closely comparable supporters.[43]

Art historians seem to agree that the latter part of Henry's reign—the second half of the 1250s in particular—witnessed "a major reorientation in the styles of English art." Not all of this, of course, can be attributed, in a tidy top-down model, to what we might call a parallel reorientation beginning to be experienced by Henry—both in his political fortunes and in his artistic interests—also in the mid-1250s. Still, certainly Henry had the means (and fervor) to articulate himself artistically, through the anything but insubstantial institution of his patronage. And overall, the evidence which documents the material dissemination of Henry's continuing tastes for certain themes (although in the context of a sea change in modes) proves by no means outside the current of stylistic change which modern scholars have identified in this period. As noted above, arguably the key event in the late transformation of Henry's decorative interests came in his 1254 "cultural expedition" to Paris, which included a visit to St. Louis's Great Hall, in the Palace of the Old Temple just outside the city. Here, "as many bucklers as the wall could hold" were hung on display, and in the anecdote which Matthew Paris relates of the incident (and which Pamela Tudor-Craig glosses persuasively), Henry appears enthralled by the scheme. "Decorative heraldry," observes Tudor-Craig, "had been an element in Henry's repertoire of motifs" since the 1240s; but immediately upon his return from France the king gave arms (at least, in their literal, standard usage) an apparently much more prominent role in his decorative schemes than ever before. Numerous examples pertain, but the point is illustrated most compellingly by a series of shields carved in 1255 in the nave at Westminster Abbey, a roll of arms, in stone, which included those of the Holy Roman Empire, St. Edward the Confessor, the Kings of England and France, and accompanying them, "just such a concourse of English nobility as had been present at the Old Temple the year before."[44]

Within the iconographic scheme of the 1260s Painted Chamber, the arms of Henry III and of Richard of Cornwall function in much the same manner as those in Henry's stone armorial at Westminster Abbey and in Louis IX's gallery of actual shields at the Old Temple: that is, each shield (a discrete iconographic unit) operates as a direct and unambiguous reference to the person of an individual (if genealogically defined) chivalric agent. In this they are not dissimilar from the arms of Richard II as they appear on the roof of Westminster Great Hall, or indeed on the Wilton Diptych (fig. 1.20), during the age of heraldry's fully codified maturity. Such usage, in fact, begins already to be apparent in Matthew Paris's 1244 *Liber Additamentorum* roll of arms (fig. 1.17, above) and in his use of overturned shields in the margins of the *Chronica Majora*, to indicate the death of their bearer (fig. 1.21). My point in this is that the wider artistic or decorative use of heraldic shields seems to be marked, from perhaps the middle of the thirteenth century, with something of the anxiety for representational stability—for a one-to-one semiotic correspondence—which was to be so widely displayed later.[45]

This distinction may be crucial to understanding the King's Westminster *mappamundi*. For if the central set of images in the 1230s Painted Chamber did not aspire to such a precise or technical representation of the official armorial achievement of England as would be employed, directly and unambiguously, to various ends in later royal decorative schemes, still it would seem that this mural scheme—especially those images which the Psalter Maps suggest may have gathered around the royal bed—appear to have operated nonetheless in the recognizably "near-heraldic" manner of the late-medieval royal emblem or badge. The device on the wall may not have been taken to be fully constitutive, or directly representative, of the individual and genealogical character of the royal knight and potential crusader sitting before it. But this map-image must have been insistently evocative of the king's peculiar authority, his mixed sacred and secular, at once chivalric and executive, feudal and fiscal public identity.

According to a heraldic reading, then, the Psalter Map's wyverns serve as its "supporters," the two censing angels act as "tenants," and key "lesser charges" appear in the wyvern-tail fleurs-de-lis, in the sets of stars (in threes) behind the world, and in the orb in Christ's left hand. Set in an azure field, the T-O map itself—a stand-in for Christ's body, as well as for the corporate community of Christian society—reads as the achievement's central "charge," becomes in other words the basic or defining feature (and perhaps to some extent it is a reserved one, a royal peculiar) of a composition which serves, ultimately, to characterize royal privilege, and to justify the unimpeded exercise of hereditary executive (as opposed to baronial communal) power.[46]

KINGSHIP, PART TWO: LEGAL AND ADMINISTRATIVE

From the combined evidence of the c.1260–65 Psalter Map miniatures and of a reconstructed thirteenth-century Westminster Painted Chamber, I have so far argued that medieval world maps emblazon royal identity; that they assert a Christological basis for royal authority; and that they tie kings to crusade obligations, if sometimes inconveniently. Henry III, as we have seen, turned away from cartographic imagery late in his reign, after embracing it early in his maturity. And though it may prove difficult to isolate the precise determining factor in this, presumably he did so (i.e., stopped commissioning world maps) for reasons beyond simply chivalric embarrassment or an increased armorial codification. In addition, he may have done so for consciously ideological reasons. For maps, and such other geographic and administrative images as Henry began to employ in the mid-1230s among the symbols of his authority, had in some of their features political and ideological implications which would help lead, ultimately, to the dismantling of that very model of a primarily sacramental or liturgical kingship which Henry came to espouse with increasing conviction and tenacity as his reign wore on, and his struggles with a constitutionally minded baronage increased. In other words, by the 1250s and 1260s there had developed a certain tension between Henry's fixed (and increasingly nostalgic) Christological view of his kingship, and that model of political organization (nascent western statehood: increasingly secularized, bureaucratized, centralized, territorialized) whose representation would come to be most powerfully embodied by newer species of those very images (territorial maps) which Henry had selected in the mid-1230s for help in "defining his position" as a ruler. For kings and for others in their audience, maps resonated differently in the later thirteenth century and afterwards, than they had in the second quarter of the century when Henry commissioned a series of cartographically oriented images for key sites like Westminster and Winchester.[47]

To grasp this, it is crucial to recognize that if St. Louis of France served as Henry's royal exemplar beginning in perhaps the mid-1250s, in earlier years Henry had looked to the charismatic Frederick II of Hohenstaufen—"that most famous and remarkable of all medieval rulers"—as a model for his kingship. The succession crisis resulting from Frederick's death in 1250 would lead inexorably to Henry's ill-fated "Sicilian Business," as discussed above; but in the quarter-century before this Frederick ruled as Norman King of Sicily (as had his maternal grandfather Roger II); as King of the Germans and Holy Roman Emperor (as had his paternal grandfather Frederick I Barbarossa); and even, for a time, as King of Jerusalem (through his second wife Isabella of Brienne, d.1234).[48] Not least important for this argument, Frederick was Henry III's own brother-in-law (via Isabella of England) from 1235–1241—key years for the emperor as they were for the development of royal cartography in England.

Frederick II has long stood out to modern scholarship for his profound impact upon the political conventions of his period, and for his close involvement with intellectual production across a number of disciplines (especially the legal and the scientific) as well as cultural divides (especially that between the Latin Christian and the Arabic). The "heroic dimensions" of Ernst Kantorowicz's seminal (pre-World War II) treatment of Frederick as "the epigone of modernization" have been trimmed somewhat by scholars in more recent years; Kantorowicz has been closely critiqued, for example, on the grounds of his arguably proto-Nazi historiography in *Kaiser Friedrich der Zweite*, a popular biography of the emperor much admired by Hermann Goering. Nevertheless, commentators seem to agree that from his extravagant and intellectually vibrant court, based first in Sicily and later in Apulia, Frederick "created an atmosphere which affected all who breathed it." The key element in this atmosphere, especially in the decade or so following Frederick's first excommunication in 1227, appears to have been the emperor and his advisors' ambitious attempt (Friedrich Heer's impression is of political ideology "being hammered out with cold deliberation") to fashion a new philosophical basis, and indeed legal justification, for the exercise of secular authority: one nominally inclusive of but in most respects outside the realm (and reach) of the definitively religious—or more precisely, the ecclesiastical.[49]

Pope Gelasius I, at the end of the fifth century, had outlined an extremely durable "theory of the two swords" by identifying the "'two things by which the world is ruled,' the 'sacred authority of the popes' (*sacrata pontificum auctoritas*) and the 'power of the ruler' (*potestas regalis*)." As Holy Roman Emperor from 1220 and thus by title as well as territory hegemonic among western monarchs, Frederick held claim to an overarching "temporal" authority in Christendom, but one which (following Gelasian theory as glossed by Carolingian imperial experience) technically was complementary to, bound up together and allied with, the pope's analogous "spiritual" authority over this same universal community. The two were, in theory, joint Vicars of Christ on earth. In actual practice, however, the empire and the papacy spent the second quarter of the thirteenth century locked in a bitter dispute for primacy in the west—one which had numerous fronts, including the military, the economic, and not least the ideological. To Matthew Paris, this dispute ranked as central among European events, in importance second only to the crusade (though by no means were the two affairs unconnected). "Indeed," as Paris remarks as late as 1247 in the *Chronica Majora*, "those of us who have read the succession of historical annals have found no other instance of such intense and inexorable hatred as existed between the lord pope and Frederick."[50]

According to Friedrich Heer's analysis of Hohenstaufen imperial ideology, "in his own interpretation of his office, the emperor was the supreme defender and protector of the church." However, "church" here denotes not so much that institution whose head was at Rome as rather "a single Christendom" in which "emperor and pope reign side by side as equals." Hohenstaufen propaganda, that is, while championing the basic concept of secular (as against ecclesiastical) authority, nonetheless in seeking to do so "set out to demonstrate that God and the world belonged together," that *regnum* ("kingdom") and *sacerdotum* ("priesthood") were not separable, competing, or hierarchical terms, but sides of the same coin.[51]

This was incendiary politics in that by the High Middle Ages the Gelasian theory of the two swords had, in the hands of the papacy and its advocates, been "transformed"—largely through a gradual accrual of Cluniac notions about the primacy of priestly over secular authority—into a doctrine maintaining "the supremacy of papacy over empire." Buttressed in particular by Bernard of Clairvaux (d.1153)'s exaltation of the pope as "God's unique representative," the papacy had, by the advent of Frederick II, effectively "destroyed the old unity of the Church as a Christendom in which pope and emperor ruled jointly and side by side, in favor of an ecclesiastical monarchy" or *regnum ecclesiasticum*. This meant that the primary political objectives which popes like Gregory IX (1227–41) and Innocent IV (1243–54) inherited with the office—most importantly, to consolidate their sovereignty in the papal states and Italy in order to ensure the continuing separation of the German Empire from the Kingdom of Sicily—derived quite directly from a developing principle of the papacy's *plenitudo potestis* (literally, "plenitude of power"). This principle "claimed the kings of the nations as vassals of the pope and the empire as his to dispose of."[52]

In the long term, this claim to world dominion had the opposite of its intended effect. For one thing, the papacy's triumph over the empire (which effectively collapsed after the death of Frederick II in 1250) came at the price of what might be called an overtaxing (during the long propaganda war) of Rome's "spiritual resources." More importantly, the papacy's "crusades to desacralize the empire" helped prepare the way for the similar "desacralization" of all Europe's monarchies, a step essential to their development into autonomous secular states. One general result of the conflict defining the second quarter of the thirteenth century, then, was that "all Europe," its intellectual as well as its political elite, "was profoundly affected by the firm resistance to an omnipotent papacy which the secular state put up under [Frederick's] leadership." The papacy prevailed (and made its point by exterminating Frederick's line) but key terms in the long-running medieval contest between the temporal and the spiritual—concepts like *regnum* and *ecclesia* themselves, and as we shall see their accompanying (carto)graphic manifestations—underwent some important modification in the process.[53]

Frederick's answer to the popes' claim to be God's unique representative has been described as an "imperial messianism" whose definitive feature was that "the emperor is appointed directly by God." This sacralized the empire, made it "a divine institution," but did so, crucially, without requiring ratification by the church (and thus acceptance of client or vassal status) as was the case in consecration-centered liturgical or Christological kingship.[54]

The principle of the empire's overall and unmediated sacralization was central to Frederick's program of imperial messianism as well as a focal point, to be sure, of papal anti-Hohenstaufen invective. Still, it bears observing that it was less in this attempt to link imperial authority to God as in just the opposite—the classical side of the coin, the Roman empire's historically and fundamentally secular character—that the contest with ecclesiastical Rome might be won (and would be, over time, by national monarchies in England and France). For Frederick to succeed in his struggle for sovereignty in the west he needed ultimately to establish that his authority, while accepting sponsorship from the spiritual realm, nonetheless derived practically from and enjoyed *de facto* dominance within an essentially church-independent secular realm, a civic sphere, whose boundaries were marked out by a set of governmental practices which in some sense could be said to contain their own justification.[55]

Many aspects of public life in Sicily and the Empire display signs of having been touched by the emperor's enthusiasm for this political and cultural initiative.[56] Had we no constraints of space, we might profitably explore the roles played by Augustan imperial ideology, and by what it appears were official attempts to promote the development of cartography, in Frederick's program of state articulation. To appreciate the implications of Frederick II's attempt to set himself up as "a new Augustus," who would renew the vigor as well as the bounds of classical empire, it is important to recognize the pronounced "geographical orientation" of Rome under Augustus, especially in terms of its political expression. As demonstrated in Claude Nicolet's *Space, Geography and Politics in the Early Roman Empire* (1991), the 'old' Augustus (63 BCE-14 CE) displayed what amounts to an "obsession with space," i.e., with the means and forms of geographic and cartographic description, especially as related to the project of articulating universal empire. Augustan Rome's marked spatial orientation is a key issue for medieval secular constructions of geography insofar as the later period's most ambitious ruling dynasties sought invariably—from Charlemagne and the Carolingians, to Frederick II of Sicily, to Edward I of England—to characterize themselves as inheritors of the imperial tradition established by Augustus.[57]

The matter of actual cartographic production in Sicily and the Empire during the second quarter of the thirteenth century must remain beyond the bounds of this chapter. I cannot bear, however, to ignore an hypothesis forwarded by H. C. Friesleben in 1984, which speculates that portolan charts (sailing charts of empirical basis and everyday use whose early history remains shrouded in uncertainty—the period's second key cartographic stream, after *mappaemundi*) may have originated as a result of the "active," "lively" and "imaginative" maritime interests of none other than Frederick II himself, and of the "navigational enthusiasm" of his Sicilian-Hohenstaufen dynasty overall, with its unique and dynamic Arabic cartographic inheritance.[58] A second hypothesis concerning imperial cartography, likewise beyond our central thrust but worth citing, involves the Ebsdorf Map, not in the context of its well-documented ecclesiastical links and meanings (how it works "to represent God") but rather in terms of an intervention this document appears to make into territorial politics. According to Armin Wolf's arguments, the Ebsdorf Map may have been produced to commemorate Frederick II's creation of a new Duchy in Lower Saxony, for the nephew (Otto the Child) of his old rival Otto IV—an act marking a legal settlement and diplomatic reconciliation between Welf (Guelph) and Staufen (Ghibelline) imperial factions. This diplomatic development was expressed, in the detail important to us here, via Frederick's opening of new relations with the English royal house (e.g., his contracted marriage to Henry III's sister Isabella), meaning that there may be imperial underpinnings to England's standing in the thirteenth century as special site for the production of large-scale *mappaemundi*—especially insofar as some of these appear to have been deployed as a new kind of royal emblem.[59]

Ultimately, the key point for this discussion as regards Frederick II is that his vision of an imperial rulership—though it was thoroughly infused with ecclesiastical concepts—had in an important sense moved past (or at least "no longer depended on") the basic principle of a "Christ-centered kingship." Instead, the arguments developed to support the principle of an extra-ecclesiastical basis for secular authority derived from the increasingly sacralized realm of law—Roman law, in particular. For the first time, under Frederick II, an essentially "law-centered" and bureaucratically oriented ideology of kingship begins, if slowly and in fits and starts, "to dissolve the image" (or "to supercede the stratum") of "the mystery-like Christomimesis prevalent in the earlier centuries." An authority based primarily in the administrative realm itself, that is, begins in the thirteenth century to displace the concept and claim of an authority based ultimately in a king's resemblance to Christ (and in his anointment by the pope or his representative)—and does so increasingly as the century progresses.[60]

A wide variety of political technologies and discourses, and of social-historical factors, are of course involved in any process of state organization, such as Frederick's Holy Roman Empire undertook (abortively) and as England later underwent. It is a key strand of my argument, however, that in the course of the thirteenth century cartography, in its various and complementary practices, was to emerge as a key factor in the battle of an increasingly centralized and bureaucratized territorial, monarchical state against what have been called the "disintegrating" and "centrifugal" tendencies of feudalism, as well as against the internationalist, theocratic claims of a universal church.[61]

This secular political potential in cartography is something Frederick II apparently recognized. But in the first half of the century, cartographic technology was not yet developed enough (despite the emperor's apparent push to get it there) to underwrite the authority of a centralized territorial administration by providing it with visible geographic inscription, an explicitly spatial and textual articulation to mark and accompany institutional procedure. Without the technology of the regnal (as opposed to global) territorial map, medieval secular administrative practice could have as yet no persuasive graphic reality (as did an internationalist, universal *ecclesia* in the T-O *mappamundi*)—the kind of sweeping, at-a-glance epistemological expression which, as the work of leading map theorist J. B. Harley has indicated, is essential to the consolidation and to the implementation, the social expression, of coercive state power. The means to such an expression Frederick II seems to have actively pursued. But the conceptual breakthroughs and technical advances in terrestrial mapping that would be necessary to realize it did not fully develop until well into the reign of Edward I, in England.[62]

Henry III's marriage to Eleanor of Provence (January 1236) ranks high, by all accounts, among the factors that in the late 1230s spurred him to a period of decorative activity among his most innovative and prolific. But his sister's marriage to Frederick II the year before may have played an even more important role in Henry's decorative activity, in particular his development of an iconography appropriate to his vision of kingship—one which may have matured in the course of extended conversations about his affairs with the emperor's marriage proxy and closest counselor, Peter de Vinea. Thomas Beaumont James has suggested that the emperor "probably provided the impetus" for, among other new expressions of royal privilege and identity, Henry's assembly at the Tower of London of a live menagerie of exotic animals. If so, it may not be unrelated—given additionally Frederick's stature as a zoologist—that in 1237 Henry ordered a bestiary to be added to the Painted Chamber, whose pictorial inventory of fabulous creatures may have been substantial already (on account of the 1236 *mappamundi*) or if not, was fit to be increased. Since bestiaries were a means of showing the nature of God "by unveiling the correspondences between the phenomenal world of nature and its heavenly archetype"—and moreover were another subject appropriate to an encyclopedic format—their frequent proximity to *mappaemundi* (in both codicological and architectural settings) may be taken as more than coincidence.[63]

It is not surprising that Frederick II should have encouraged Henry to "assert his own independence of papal policy," as Maurice Powicke has reported. For his part, Henry's sensitivity to his new imperial ties and his inclination to regard Frederick and himself as "allies in the direction of the affairs of Western Christendom" meant that he can hardly have been unaware of the emperor's innovative conception of his own rulership, and of his administrative re-organization and legal re-imagination of the territory he ruled.[64]

The choice for Westminster in 1236, and for Winchester again in 1239, of the icon of the world map, as a vehicle appropriate for the expression of his nature as a secular ruler, has undeniably important sacramental implications. However, this iconographic choice on Henry's part has, simultaneous to these, certain other political implications which depart significantly from the principles of Christological kingship. These elements are to be found in such a map's connections—both thematic and diplomatic—to Frederick II and the momentum of his imperial project. The rest of this chapter will look to examine the particular hybridity of this new kind of royal emblem, the world map that is at once Incarnate Word and heraldic shield, that is both Body of Christ and document of imperial administration.

PAINTED CHAMBER *COMPILATIO*

The introduction in the early 1260s of the two world map miniatures analyzed above (figs. 1.7 and 1.8) onto the front of a London Psalter, British Library MS Additional 28681, has typically seemed worthy of mention to commentators on the history of English cartography; but usually just in passing. At one point there flourished some disagreement among specialists regarding whether the leaf bearing maps on its recto and verso (now folio 9) was produced independently, then affixed to the Psalter for a frontispiece (as is the case for the illuminations covering folios 1–8, added c.1300), or should be regarded instead as having been integral to the codex all along. In any case—latterly affixed or (as now accepted) not—these two miniatures' presence within a Psalter, the century's most common "picture book," often has been implied to be a codicological act more or less serendipitous, even mistaken, from the standpoint of usual practices in medieval iconographic compilation. What does a monumental, encyclopedic, classically derived and heraldically inflected world map like King Henry's have to do, after all, with a Psalter—in the twelfth century an object, it is true, of public liturgical use (appropriate as a gift from lay patron to religious community) but in the thirteenth, increasingly a book of private devotion, not unlike the small-dimensioned Book of Hours which was developing out of it?[65]

Far from being mistaken, Additional 28681's incorporation of a leaf bearing two Christological world maps makes considerable codicological, not to mention thematic, sense. For even if we have no explicit manuscript precedent for this striking incident in liturgical and/or devotional illumination—no other known Psalter features a prominent, actual map—we do at least have an architectural parallel: Westminster's 1230s-1260s Painted Chamber. This parallel provides the basis for the reading presented in the remainder of this chapter.

By the twelfth century English Psalters had developed, as a result of their use in the liturgy, a standard ten-part division marked by specially ornamented or "historiated" initials at the places in the text at which daily reading would begin: Psalms 1, 26, 38, 51, 52, 68, 80, 97, 101 and 109. Each division had its traditional iconographic subject (e.g., Psalm 1, King David as Psalmist; Psalm 109, The Trinity), which in luxury editions might cover a good deal of the page. More importantly, ornamentation within the text proper (i.e., borders, initials, line endings, marginalia) was often supplemented by a prefatory (though sometimes also interspersed) set of illuminated miniatures (as in the Map Psalter's added ff.l-8), treating scenes from the Old or New Testament including among others the Tree of Jesse, the Virgin and Child, and most often of all Christ in Majesty. Although differences between versions can be many, in this last subject Christ frequently sat enthroned, right hand blessing and left bearing orb or book, in a quatrefoil surrounded by the Four Evangelists or their symbols—an arrangement attested by two c.1250–60 London or Westminster manuscripts which have been linked to the Map Psalter (the single-leaf Cleveland Museum of Art Acc. 45.132 and the Rutland Psalter, B.L. MS Additional 62925, fol. 112v; see fig. 1.22). Set prior to the text of the Psalms in a great many manuscripts (including the Rutland and Map Psalters) was a Calendar, noting feast days, saints' days and occasionally patron obits, hierarchically by ink-type and with anywhere from one to six months per manuscript page. Standard format also included (after the Psalms) the Canticles and the Litany, and sometimes additionally the Hours of the Virgin, the Office of the Dead, and/or other texts of a liturgical or devotional character.[66]

The traffic I am looking to establish between mural-subject selection for the Painted Chamber and that for miniatures in contemporary illuminated Psalters may be more easily seen for some icons than for others. It will be recalled that Henry III commissioned a Calendar for the Painted Chamber in 1236 and in 1243 ordered the Four Evangelists added, one per wall (as each went to a corner of world or map), followed in 1259 by a Tree of Jesse above the fireplace. All these, as we have seen, are set subjects in mid-century Psalter illumination, with indeed London or Westminster manuscript witness (the Map Psalter's calendar, like its frontispiece map-leaf, has been judged original to the compilation). Similarly, the initial scheme's mysterious *magna historia* or "great history" (if this does not constitute reference to the map) would appear to have its own contemporary and local Psalter counterpart, in the standard set of prefatory scenes from Christian history. A painted bestiary—*panellis continentibus species et figuras leonum, avium et aliarum bestiarum* ("panels containing the forms and figures of lions, birds, and other beasts")—though ordered to be left off in 1237 so as not to impair the effect of the *magna historia*, had also been included initially. The bestiary has direct links to the world map, as we have seen, and appears, like it, to have been "especially characteristic" both of England and of the thirteenth century. But the bestiary also connects with the Psalter in that illuminators who decorated manuscripts of the one were usually involved likewise in production of the other. The reputation of the above-noted Rutland Psalter, for example, rests largely on its marginal grotesques, a series which includes numerous scenes adopted from the *Marvels of the East*, a bestiary variant connected with the travels of Alexander the Great.[67]

The remaining and apparently central image in the 1230s-60s Painted Chamber, of course, was that peculiar version of Christ in Majesty with world map an appended part of, or held conspicuously before him—this according to the testimony of the Psalter Map miniatures, at least. Setting aside the feature responsible for their uniqueness (i.e., their outsized maps), parallels otherwise abound between these illustrations and contemporary English Psalter versions of Christ in Majesty, as well as some renditions of the miniature-subject for Psalm 109, The Trinity. (These subjects link together textually and iconographically: in the Rutland Psalter, for example, folio 113r's Psalm 109, with initial depicting The Trinity, lies directly opposite folio 112v's Christ in Majesty.) In fact, using elements taken from images and initials found within contemporary Psalter illumination, a composite might very easily be built apprehending most of the Psalter Map-miniatures' non-cartographic features. No exact parallel to the specific combination of pictorial elements appearing in the Map Psalter illustrations now survives (or should be expected to), but to trawl through a series of associated productions will illustrate how the sibling compositions most central to this argument effectively amount to amalgamations; they are constructed, that is, of iconic elements drawn from contemporary illuminatorial practice. Notable parallels to Psalter iconography may be found, for example, in the presence of Christ and/or God enthroned; right hands blessing, left hands bearing orb; "Veronica"-like heads; tenant angels censing; dragons underfoot, indeed with fleur-de-lis tail flourishes; and blue backgrounds set with white stars in clusters of three—all of this encased, typically, by floral border decoration. The T-O orb in enthroned Christ's left hand (on the recto Psalter Map and elsewhere; see figs. 1.7 and 1.23) is of particular cartographic interest, perhaps the more so in that other versions of Christ in Majesty—e.g., the Cleveland Museum leaf and twelfth-century French church-facade sculpture almost invariably—fill this hand instead with the Book of Life. It may be that thirteenth-century England's "geographical culture" saw a special opportunity in the introduction of the T-O icon, whether torso *mappamundi* or hand-held globe, into Christ enthroned in Majesty. This culminating image both of theological history and in Psalter miniature-sets is thus made more cosmographically encapsulating.[68]

This point of a correlation between mid-century Psalter iconography and the murals selected for the 1230s-60s Painted Chamber need not be pressed too hard. What is noteworthy is that such overlap (and its attendant wrapping of kingship in piety) becomes more apparent—with the addition first of

Evangelists (1243), then of Jesse Tree (1259)—in years which saw Henry's privileges and authorized standing as a sacramental king most eroding to the forces of baronial communalism. After Henry's failed Poitevin expedition of 1242, Simon de Montfort, the king's brother-in-law but a voice already for disgruntled baronial interests, told him "in words which Henry never forgot . . . that he deserved to be shut up like Charles the Simple," the tenth-century King of France who had been imprisoned and died in captivity. If Henry was "not again led astray" in an imperially inflected "continental adventure" until the mid-1250s' Sicilian Business, this time baronial opposition galvanized to effect practically that: the Provisions of Oxford (1258) left Henry more or less a figurehead. It will be recalled that from 1244–59 the Painted Chamber likely saw some early meetings of Parliament. This room had long been crowded with images. But the addition of just a few more, above the fireplace or high up the wall, may have underlined the Painted Chamber's Psalter association, its expression of pious or liturgical kingship. Such fits well with the predominantly sacramental themes which in these years were being developed more fully elsewhere, at less crowded or newly renovated royal decorative sites including Westminster Abbey in particular—where the royal-sacral figure of St. Edward the Confessor came more and more to dominate.[69]

Yet Psalter illumination practices alone do not fully explain the 1230s mural scheme's curious roster of images, particularly its map (and perhaps its *magna historia*). Another medieval textual tradition has strong claim to compilatory influence as well—one characterized by an imperial Roman administrative, historical, and cartographic orientation, rather than by a pious or devotional Christian one.

Pamela Tudor-Craig has observed that, overall, medieval *mappaemundi* have a "dual context": they accompany not just written but also illustrated history. Tudor-Craig finds a parallel to the striking juxtaposition of mural subjects in the 1230s Painted Chamber not in the medieval Psalter, as I have done above, but in the *Notitia Dignitatum* (literally, "list of office-holders"), a "compilation of geographic and administrative information" which covers the western half of the Roman empire and which has been dated in its initial form to c.395. Illustrated from an early stage (c.425–30) with stylized maps, city-views, personifications of provinces, and seated figures of officeholders ("attired like kings"), the *Notitia* continued to be consulted, if less frequently copied, through to the end of the Middle Ages, where it is described on one occasion as a *mappa mundi* itself. In the Renaissance, when interest in the text revives considerably, each copy owes "to a patron particularly interested in classical antiquity." So much is hardly surprising, as even the fifth century's "creation of a *de luxe* copy" (i.e., an illustrated one) has been put down to the influence provided by "a period of renewed Augustan classicism in the arts in Rome"—"just the right context."[70]

The *Notitia Dignitatum*, by most accounts, served as a kind of report on the "condition and resources" of the Western empire. In essence it is a record of imperial postings *tam civilium quam militarium* ("civil as well as military"), as the full title of the document runs, and it acts as an elaborate flow-chart, as it were, of imperial responsibility, authority, and official precedence. In layout, the document proceeds geographically and charts the "hierarchical relationship" of all officeholders to "centralized bureaucracy," a bureaucracy which is personified in its highest ranking provincial figures (or rather, their offices). To this end the *Notitia* sets out insignia of office (i.e., description and/or illustration of badge, emblem, or mark; see fig. 1.24, a map of Britain with castle-icons marking its five provinces); followed by information as to a given office's "sphere of competence," including a hierarchical list of subordinate offices; followed finally by a list of "immediate office staff attached to the position." An important source for Roman Britain, the *Notitia* has been studied mostly for what it reveals of late imperial bureaucracy and only very rarely in a medieval context.[71]

All extant versions of the *Notitia* derive from a lost exemplar produced at Speyer (Germany) in 1436, an exemplar copied itself not from a late-antique manuscript but rather from an intermediary Carolingian version. The classical imperial inheritance claimed by the Carolingian Emperors, as J. J. G. Alexander has said, "would have provided the necessary motives for the copying of the text and the pictures." But what is more, this act connects us again with the will of Charlemagne (d. 814) and its pregnant reference to three silver tables engraved with the representation (*descriptio*), respectively, of Constantinople, Rome, and the "entire universe" or "whole world" (*totius mundi*) in concentric circles: could these, Alexander asks, "have derived from an illustrated *Notitia*?"[72]

If each reproduction of the *Notitia*—whether fifth, ninth, fifteenth-century or later—served, in essence, to mobilize classical history in support of an incumbent secular power with imperial pretensions, such a text should likewise have been of use to an English monarchy in the thirteenth century. No copies are extant from England, even late ones; but speculation since 1552 has held that the *Notitia* was recovered (or recoverable) *ex ultimis Brittanis* ("from out of Britain"). The transmission between Carolingian and Speyer versions, that is, appears to have involved or passed through England. C. E. Stevens, the strongest proponent of Britain's claim to medieval acquaintance with the *Notitia*, holds categorically that "a copy was in England in 1200." Giraldus Cambriensis (c.1146–1223)—author of a number of historical-geographical works on Britain and a noted cartographer himself—worked on at least one occasion from a manuscript, Stevens argues, which included the *Notitia Dignitatum* (as well as the *Notitia Galliarum*, a related text which treats ecclesiastical Gaul and is extant in three English copies), a manuscript he "presumably" found at Rome. Stevens argues further that Arthurian chronicler Geoffrey of Monmouth (*Historia*

Regum Britanniae, c.1136) had access to just such a text and that "he, at least, must have found it somewhere on this island." Other commentators note a likely sounding "cosmographical codex" mentioned by Bede (d.735), and another which Bishop Lullus of Mainz "tried to get from York" c.770. But Stevens speculates that "it ought to have been Alcuin"—Charlemagne's English tutor and advisor, "the most learned man anywhere to be found" in the words of Einhard—"who sent over the volume which Geoffrey and Giraldus were centuries later to read." Alcuin was the pupil of Bede according to Notker the Stammerer (although in fact it is Alcuin's master Egbert whom Bede taught); in any case Alcuin appears to have known the *Notitia* and appropriately enough Einhard reports that he focused his efforts with the emperor especially on astrology.[73]

Speculation aside, British Library MS Cotton Tiberius B.V, an eleventh-century manuscript of asserted London provenance and possible royal ownership, provides a documentable "medieval parallel" to the late-classical *Notitia*. This compilation's last and key item is an imperial travel and bestiary-text, the *Marvels of the East*. Prior to the *Marvels*, the manuscript contains an illustrated Calendar; lists of officeholders including Popes, Emperors, and English Bishops and Kings; astrological and astronomical texts and pictures, including a planisphere and other diagrams; together (among much else) with two *mappaemundi*. One of these is circular but the other is the well-known, rectangular Cotton/Anglo-Saxon world map, whose straight-lined demarcation of Roman provincial boundaries is prominent as on no other medieval map (see fig. 1.25). Another medieval English parallel to the *Notitia Dignitatum* is provided (although the *Notitia Galliarum* may offer the closer comparison) by Gervase of Canterbury's c.1262 *Mappa Mundi*—a text appended to his *Gesta Regum* ("Deeds of Kings") which is neither "map" nor "of the world" but rather a three-column list, covering Britain only, of ecclesiastical and civic administrative divisions.[74]

If the *Notitia Dignitatum*'s apparent links to Geoffrey of Monmouth and Gerald of Wales begin to place this material and its themes (Roman imperial; administrative; cartographic) in the context of narratives of English kingship (Geoffrey) and British Island territoriality (Gerald), a more pointed connection to the matter at hand (i.e., a Painted Chamber map) comes via the thirteenth century's preeminent handler of these subjects whether verbally or visually—we are back to Matthew Paris of course—some of whose links to royal decorative sites we noted above. For Westminster connects with Paris and his St. Albans scriptorium on precisely the subject of what has been called the "evolution of a quasi-classical iconography of English kingship." "Quasi-classical" maps and the administrative images and texts related to them, in other words—as exemplified in the *Notitia Dignitatum* but also as suggested by the politics and patronage of

Emperor Frederick II—serve to conjoin Matthew Paris's chronicles with Henry III's buildings, and in doing so show themselves to occupy a position central to thirteenth-century English representations of kingship.[75]

Windsor Castle's painted line of seated kings, for example—a fragment of which is the only survivor from a great deal of wall-painting commissioned by Henry in the 1240s—has direct links to the genealogical line of English royalty drawn c.1250 in a manuscript of Paris's *Historia Anglorum* (fig. 1.26). If 1259's Tree of Jesse—such images depict Christ's royal genealogy, ascending from Jesse through the prototypical King David to the Virgin Mary—strengthens a Psalter hence sacramental association, it also underlines, simultaneously, the Painted Chamber world map's quasi-heraldic function, its figuring of royal privilege through collective hereditary identity. Such images—this and lines of office-holders, as at Windsor Castle, in Paris's *Historia Anglorum*, and in the *Notitia Dignitatum*—help ground contemporary kingship with the weight of an authorizing genealogy, one which is implied to span biblical, classical, and English regnal history alike. Further, in that each figure bears in hand either an architectural detail (recall those on M. Paris's itinerary map; fig. 1.4, above) or an imperial orb (reputed to contain soil from the furthest reaches of the earth), a line of kings reinforces secular officeholding's historically administrative nature (i.e., it follows the model of classical empire), and gives this not just graphic but decidedly geographic expression.[76]

Circa 395, a geographic list of administrative divisions and a line of office-holders may have fulfilled a directly bureaucratic function. But most of a millennium later such a document had become largely fetishized, serving now as a citation of classical precedent (meant to complement Biblical typology) in the symbolic expression of authorization for secular dominion. Rome's cultural and political authority, not to mention its model of wide geographic sway, had much appeal for England's imperial-eyed monarchy—a point which in Edward I's reign we will see more clearly.

Like other texts, maps "may contain multiple, even contradictory meanings"—a point I believe it is absolutely crucial to stress.[77] A *Notitia Dignitatum* context to Henry III's lost Westminster *mappamundi* need not therefore be seen as necessarily incongruous with, nor should it be imagined to operate exclusive of, the Psalter context introduced by MS Additional 28681. The implications of both these textual traditions, as too those of the heraldic context that may have inflected the composition, are essential to interpretation of the Painted Chamber's iconographic grammar, its multilevel and mixed sacred and secular pictorial program.

Above we cited Psalter parallels for a greater part of the Psalter Map miniatures' non-cartographic or marginal features. But in their cartography *per se* these miniatures display, in turn, their concomitant *Notitia Dignitatum* context—to the extent at least that the *Notitia* operates as an administrative itinerary, a procession across the provinces and down through the ranks of imperial bureaucracy, which is, despite annotative illustration, essentially a textual or verbal more than a graphic or figural one.

We have looked briefly at some details in the frames around the Psalter Map miniatures, and these are icons, clearly, of major symbolic import. It is noteworthy that the two map-images themselves, however—and certainly space is a factor—in general avoid (the recto for the most part, the verso almost completely) the usual *mappamundi* practice of presenting copious material of a strictly speaking non-geographic nature. Largely absent, in other words, is the encyclopedic and at times "fantastical" visual and textual information (biblical, classical, legendary, zoological) for which, to many, medieval world maps are most noteworthy—Lot's Wife, Tower of Babel, Labyrinth on Crete, Polar Bear in Scandinavia, and so forth.[78] What the recto map provides instead (though there are traces of such iconic practice, monstrous races and Garden of Eden most notably) is primarily an in-curving distribution of place-names upon a topographic canvas of note mostly for its attention to waterway details.

Where the recto map's dominant spatial mode is concentric (on which more in a moment), the verso map for its part cannot be described otherwise than by its aggressively tripartite geometry. Its T-lines and O-lines (figuring continents) are absolute and clean, not jagged or convoluted as on other maps (including the recto). And beyond this there is no topography, no shape at all—as if to show that conceptual reality (that of administrative and scholarly division) underlies and may indeed take precedent over the variability of topographic manifestation. For within the bounds of each of this map's three angular continental sectors the concept of relative space is abandoned entirely. So too is iconic representation.

In short, the method of arrangement of geographic information shifts (when we move from outside to inside the map-image) from a mode which is spatial and iconic to one which is starkly linear and textual. The organization of data within the verso T-O diagram proper follows the left-right, top-bottom, line-to-line progression characteristic of standard scribal practice in the medieval Latin west; it follows *codio*graphical not *carto*graphical *mise en page*, so to speak. Thus this map's geography is illustrated but also textual in the way that the *Notitia Dignitatum*'s seems to be. Within the clear spatial frame of each continent (as within each manuscript section, divided by province and marked by cartographic insignia) what is set out is effectively a list of administrative divisions: for the most part imperial provinces like on the Cotton/Anglo-Saxon Map, though there are exceptions. Each list of names begins in its continent's upper-left-hand corner, then runs across and down some fifteen lines to end in the lower right of this pre-described space. Their most striking feature is how—

in continuing, for example, from *Bulga* at the end of *Europa*'s line 8 (where it collides with the T-stem usually marking the Mediterranean) to *ria* at the beginning of line 9—geography is inscribed upon this image more after the habit of a common scribe than after that of an illuminator, a painter of walls or panels, or what would come to be called a cartographer. For even where large-scale world maps (e.g., Hereford, Ebsdorf, Vercelli) do themselves normally include a great deal of textual material—as Woodward has pointed out, *mappaemundi* were "at least as much written as drawn" (see fig. 1.27)—their arrangement is yet much more spatial in conception. Each text, discrete, is a kind of legend placed beside the icon to which it serves as gloss. But such encyclopedic intertexts as these the recto and verso Psalter Maps have either no room for, or no inclination to include.[79]

In fact, contemporary developments in practices relating to manuscript production and organization appear to have played a definitive role in the Psalter Map-miniatures' unique mode of mixing the visual and the textual in their representation of world geography. Malcolm Parkes has documented the influence of the concepts of *ordinatio* and *compilatio* on book-production in the thirteenth century, a period he terms "the age of the compiler" and one in which the demands of an increasingly rigorous and scientific academic readership brought issues relating to the presentation of texts under new scrutiny. An increased focus in academic discussion on "precise definition" compelled readers to take a close interest in their texts' particular modes of "organization and procedure," Parkes notes. Such attention on a practical level led to formulation on a theoretical level of the concept of *ordinatio* (which is to say, a given argument's progression and precise arrangement into divisions), with the result that in time the "structure of reasoning" characteristic of the period saw its demands for close knowledge of an argument's divisions "reflected in the physical appearance of books." Numerous aids to reference were introduced as guides to the *ordinatio* of a work and to enable efficient reader-access (e.g., running-titles; rubrics marking book, chapter, and subchapter-divisions; line and column numbers; underlining of key words; use of several inks and letter-sizes; etc.). But such a scholarly apparatus was developed not just for use in new or newly copied texts. Guides to the *ordinatio* of earlier texts needed also to be produced, to bring old books into line "with new ways of thinking."[80]

The drive to make inherited material thus available to contemporary intellectual needs, through the systematic imposition of a new *ordinatio*, led in the thirteenth century to a changed deployment of the scholarly concept of *compilatio*. "Compilation was not new," Parkes stresses. New was its elevation into a sophisticated and ambitious literary form, one able at its highest level to, for example, "enclose natural science, Christian doctrine, and the history and achievements of the human race" within the framework of a "mirror of the universe" (this according to the scheme of Vincent of Beauvais). The compiler's role was to impose a new arrangement onto that which he extracted from others: "The *compilatio* derives its value from the authenticity of the *auctoritates* employed, but it derives its usefulness from the *ordo* in which the *auctoritates* were arranged."[81]

What enabled any compilation's usefulness, then, was on one level the page-to-page inscription upon a text of the reference apparatus noted above. But this apparatus developed also in another manner, into an independent form known as the *tabula*. Typically an alphabetical index, the *tabula* was a kind of subject concordance or guide to subordinate topics designed for use alongside, or as a digest in place of, the text whose *ordinatio* it described. *Tabulae* were sometimes bound up within the work to which they referred (hence the modern "table of contents"), but were more often extracted and bound into collections, "thus being made available for use in the context of different arguments." In practice *tabulae* provided a quick means of reference, a concise index to, or stand in for, the dilated contents of a larger work of manifest and usable authority.[82]

Throughout I have characterized the two Psalter Map-miniatures as apparent copies of a much larger mural original. And just above I have invoked the concept of a shift in representational modes within these compositions, from the iconic or figural to the textual. In recent work on thirteenth-century manuscript illumination, Suzanne Lewis has cautioned, however, that "visuality and textuality cannot be productively sustained in a relation of binary opposition: images are read, and texts are imaged," such that "the two cognitive channels rarely operate independently." With this in mind I believe we may adopt for use in our (primarily) visual context Ralph Hanna III's apt characterization of the textual compiler as a figure who in providing "connective tissue" between bits of *auctoritas* becomes "a subdued voice of cosmic statement." Just, that is, as the textual compiler is an "arranger of statements" which, joined together, highlight a "universal truth," so similarly is the iconographic compiler (whether manuscript illuminator or mural-scheme designer) an arranger of images which, joined together, serve to figure collectively some universal truth. And just as the literary compiler's work is "anchored in the unquestionable transcendent Word," as witnessed by a persistent citation of biblical text and authorized commentary, so too is the compiled visual text of the Painted Chamber and/or Map Psalter anchored in the world that is the Word, made manifest in Creation. As Lewis has said, "God authored two books"—the Bible and the universe.[83]

Henry III's 1230s-1260s Painted Chamber thus acts, in effect, as an iconographic *compilatio*, the redeployment in a new *ordinatio* of a collection of seemingly heterogeneous images but all within the context of the twin framing devices, cosmographic and architectural respectively, of *mappamundi* and king's bed-and-audience chamber. In this light the two (proposed) miniature witnesses we have to this room's contents, B.L. Additional 28681 folio 9 recto and verso, needn't be seen as conflicted copies or mutually disqualifying renditions of a larger image each was meant to reproduce line for line and portion for portion, if in radically reduced scale. Instead, we may see these Psalter Maps as twin *tabulae*, a kind of geo-exegetical index, in composite, to the existing or recently damaged monumental figure of the world on the Painted Chamber's wall.[84] And perhaps, on a second level, they refer back simultaneously to the key textual traditions of the *Notitia Dignitatum* and the English illuminated Psalter, which underlie them.

The development of *tabulae*, then, as guides to a compiled text's arrangement and contents, appears to have inflected the process by which two mid-century London book-illuminators sought to represent in tabular form (that is, to present again with a new and more concise *ordinatio*) the massive text of King Henry's monumental mural-scheme. Where the recto map-*tabula* may be read as a figural guide to the concentric mural *mappamundi*, setting out place-names, waterways, and a few icons in basic spatial relation to one another, the verso map-*tabula* is this geography's accompanying textual guide, setting out the divisions of imperial administration after the manner of the *Notitia Dignitatum*. This late-classical compilation's medieval English parallel (B.L. Cotton Tiberius B.V), it will be recalled, includes an important Roman-based world map with just such boundaries prominently marked. In this way Painted Chamber *compilatio*, as witnessed by the Psalter Map miniatures, invokes both figural and textual traditions of geography, just as it cites both sacramental and administrative conceptions of an English king's authority, encasing it all within the frame of universal geo-history. Such compilatory ambition falls not far short of Vincent of Beauvais's.

The Painted Chamber, in the speculative and reconstituted reading I offer of it, not only linked Henry to a series of symbolic Christian narratives, genealogies, and cartographies. It also tied him to a set of texts, images, and institutions, precisely those which helped form the basis for the thirteenth century's new conception of royal authority—one characterized by a vigorously secular as well as a renewed Christian imperial mode. Frank acknowledgment of an imperial administrative context for the Psalter miniatures' cartography is important because it registers a Painted Chamber map's inclination toward focus on the kind of secular and technical geographic information which would prove a hallmark of later mapping. The increasing technical orientation of late medieval terrestrial maps derives, then, not from the nautical example provided (as has frequently been asserted) by empirical portolan charts alone.

The "numerous reflections" of an imperial administrative and cosmographical tradition that are to be found within Matthew Paris's *oeuvre* have been read by some as "strong indication" that such ideas as the *Notitia Dignitatum* encapsulates were not only known to the St. Albans chronicler but "equally familiar to Henry himself."[85] If so, such ideas must have been known also to others in the royal household—for example, Prince Edward (the locus of authority in England from the mid-1260s onward) and his first cousin and seneschal Edmund of Cornwall (patron of a map produced c.1260s-1283 upon the model of the Painted Chamber *mappamundi*), both of whom grew up surrounded by the visual detritus of Henry's flamboyant reign.

The "close interrelationship" that scholars have identified "between court art and the St. Albans scriptorium"—both noteworthy for having an outstanding interest in cartographic representation—may have largely dissipated with the death of Matthew Paris in 1259, by which time Henry had apparently turned to largely French models, supplied by the pious and charismatic St. Louis, for both his art and his kingship.[86] Yet the geographic and administrative elements of the Painted Chamber's symbolic cartographic decorative scheme, its partial underpinnings in classical imperial text and image, would in a changing political and social context prove ready to recombine, not many years into the reign of Edward I, into a new kind of image. On one level, this emergent order of spatial composition appears conservative insofar as it remains a constituent part of the universal *ecclesia* (or institution of the world church) described by T-O *mappaemundi*. On another level, however, the cartographic document that will mark and most characterize the late thirteenth century proves innovative, in that it redrafts its relationship to this tenaciously centripetal metropolitan geography in a manner which was new in being specifically monarchist, increasingly state-oriented, and territorial on its own terms: regnal, not universal.

To Henry of England in the 1230s, the implications for kingship of cartography must have seemed positive and unique. A map filled not simply a sacral role as crusade talisman and Christological underwriter of sovereignty, but might also, after the still ascendant model of Frederick II, provide the means for asserting an authority independent of the papacy's claim to universal dominion. Thirty years later, however, to a monarch worn down by political failures both abroad (e.g., the Sicilian Business) and at home (e.g., the Provisions of Oxford), the situation may well have been reversed. The Painted Chamber at Westminster, during these thirty years, was decorated with what seems to have been an attempt to deploy cartography in support of the combined sacramental and administrative claims of English kingship. Given the opportunity of a fire which may itself have had links to the Barons' Wars, Henry had this mural program painted over in the mid-to-late 1260s, in favor of that visual embodiment of pious kingship, the unambiguously sacramental Coronation of St. Edward whose production marks one of the high points in intensity of the English royal cult.[87]

Figure 1.1. The Painted Chamber at Westminster, facing west; view by William Capon, c.1799; reproduced by permission of the Society of Antiquaries of London.

Figure 1.2. The Painted Chamber: Coronation of St. Edward the Confessor, c.1263–67; watercolor copy by Edward Crocker, c.1819; reproduced by permission of the Ashmolean Museum, Oxford.

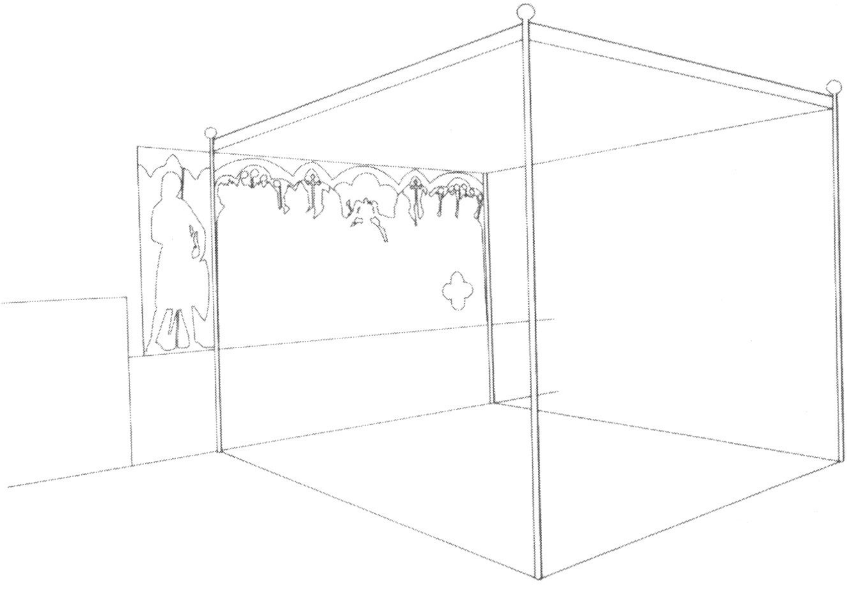

Figure 1.3. Conjectural view of the Great Bed in the Painted Chamber, showing Coronation and Soldier-Guardian murals in place, with fireplace to left; from Paul Binski, *The Painted Chamber at Westminster Palace* (London, 1986); reproduced by permission of the Society of Antiquaries of London.

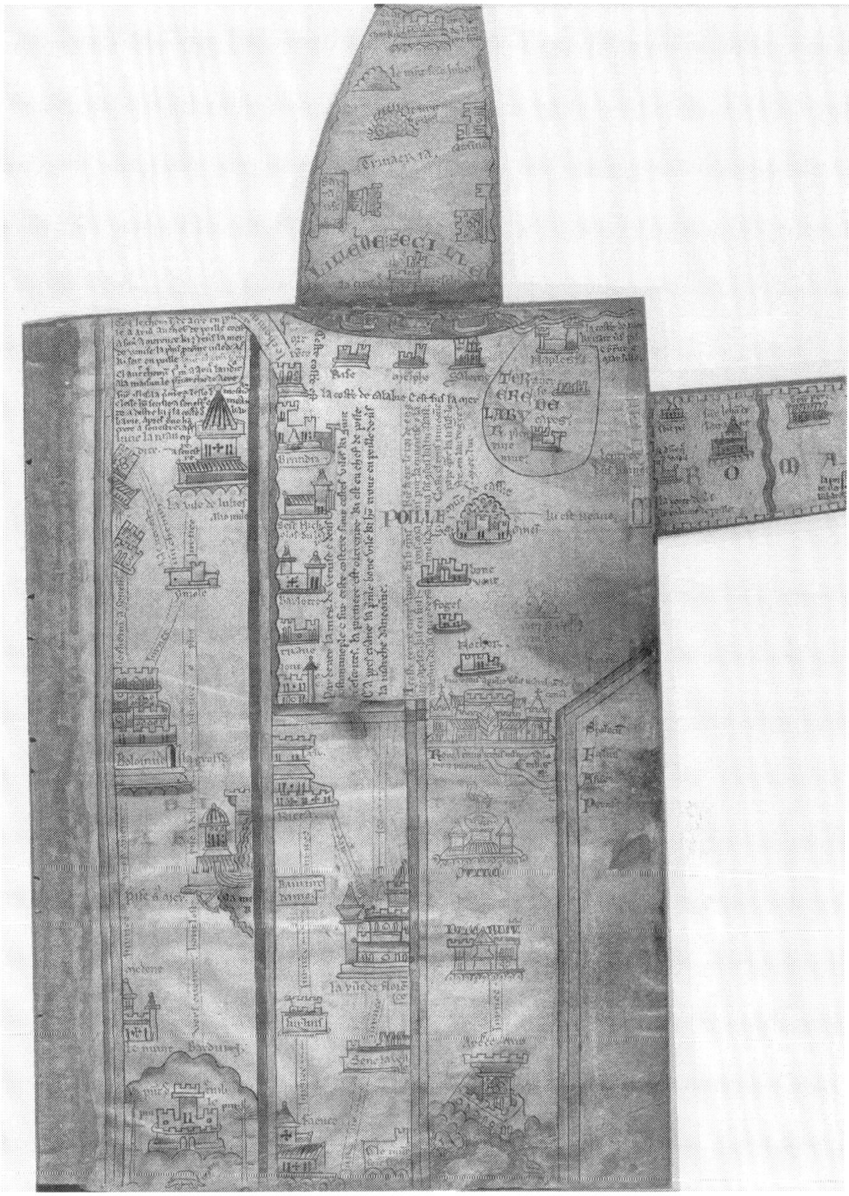

Figure 1.4. Apulia, from Matthew Paris's London-Apulia Itinerary; London, B.L. MS Royal 14.c.vii, fol. 4; c.1250–55; note the stork resting in the tower of "Sutre"; reproduced by permission of the British Library.

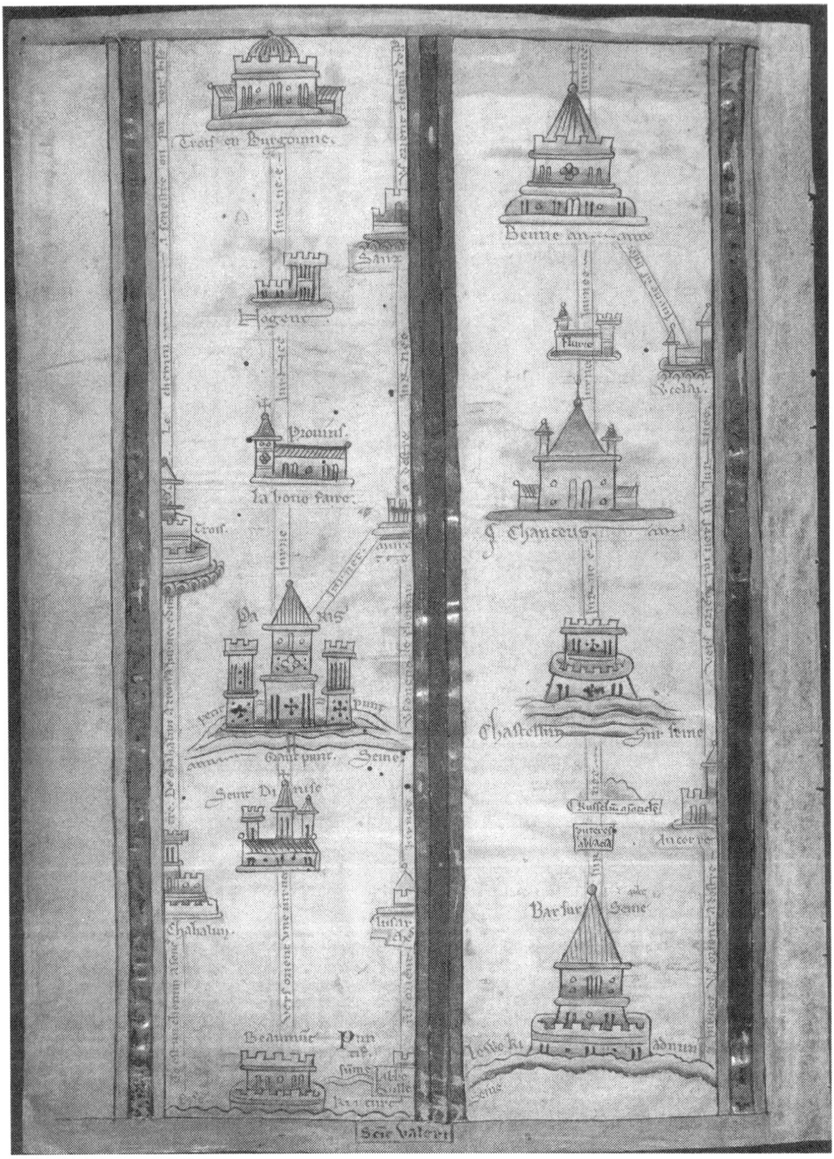

Figure 1.5. Troyes-Bar sur Seine, from Matthew Paris's London-Apulia Itinerary, London, B.L. MS Royal 14.c.vii, fol. 2v-3r; c.1250–55; reproduced by permission of the British Library.

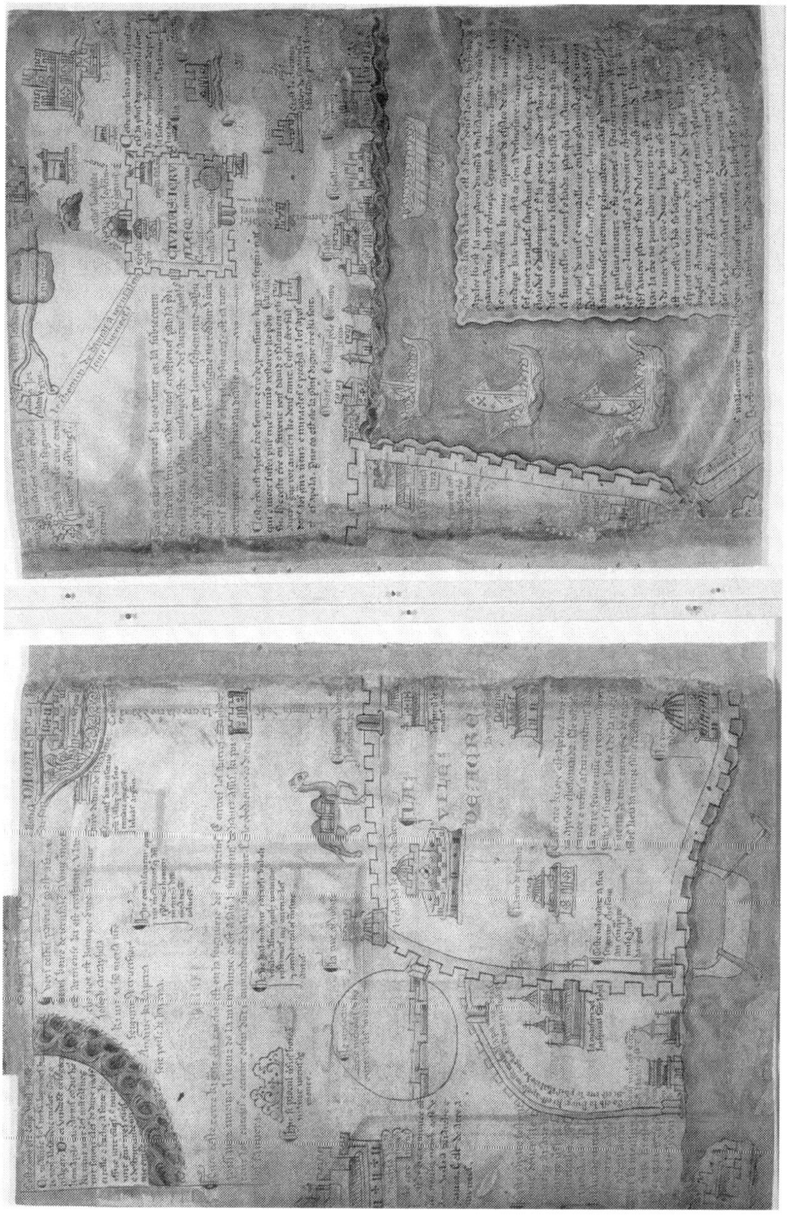

Figure 1.6. Map of Palestine, by Matthew Paris; London. B.L. MS Royal 14.c.vii, fol. 4v-5r; c. 1250–55; Acre at center-bottom, Jerusalem upper-right; note *la chemin de damas a jerusalem* (diagonally up and left from city) and *la chemin de jafes a jerusalem* (straight below it); reproduced by permission of the British Library.

Figure 1.7. The Psalter Map, Recto; London, B.L. MS Additional 28681, fol. 9r; c.1260–65; reproduced by permission of the British Library.

Figure 1.8. The Psalter Map, Verso; London, B.L. MS Additional 28681, fol. 9v; c.1260–65; reproduced by permission of the British Library.

Figure 1.9. Seal of Emperor Frederick II holding cross-surmounted T-O orb, from Matthew Paris's *Chronica Majora*; Cambridge, Corpus Christi College MS 16, fol. 127, c.1250; reproduced by permission of the Master and Fellows of Corpus Christi College, Cambridge.

Figure 1.10. Seal of Richard of Cornwall, King of the Romans, 1257; Landeshauptarchiv Koblenz, Bestand 133 Nr.18; see also London, B.L. Additional Charter 1051 (insufficiently legible for useful reproduction).

Figure 1.11. The Painted Chamber, Window Splays: the Triumph of Largesce over Covoitise and of Debonerete over Ira; c.1263–67; watercolor copies by Edward Crocker, c.1819; note prominence of Shield of England and alternation in border of the arms of Richard of Cornwall (imperial eagle) and those of Henry III (three lions passant); reproduced by permission of the Ashmolean Museum, Oxford.

Figure 1.12. "Henry of Mainz" or Sawley Map, frontispiece to a Durham-produced copy of the *Imago Mundi*; Cambridge, Corpus Christi College MS 66, fol. 2; late twelfth century; note angel supporters; reproduced by permission of the Master and Fellows of Corpus Christi College, Cambridge.

Figure 1.13. Woodcut Coat of Arms, with Angel Supporters; *Statutys for the orderi[n]ge of purveyours or achatours for the kynges moost honourable housholde*, fol.1; printed at London by Richard Pynson, 1500; compare angel supporters to those flanking "Henry of Mainz"/Sawley Map; reproduced by permission of the British Library.

Figure 1.14. Henry IV of France on the Lit d'Honneur; two engravings, early seventeenth century; Paris, B.N. MS Clarembault 1127, fol. 25v; note canopied bed and proliferation of heraldic imagery; also oval shield with fleurs-de-lis on wall behind bed, directly above body of monarch; reproduced by permission of the Bibliothèque Nationale de France.

Figure 1.15: Great Seal of Henry III, 1218 (with nineteenth-century engraving); London, British Library Cotton Charter XI.53, obverse; reproduced by permission of the British Library.

Figure 1.16: Great Seal of Henry III, 1259; Durham Cathedral MS 1.2.Reg. 6a, obverse; same matrix used for 1274 Lesser Seal of Edward I; reproduced by permission of the Dean and Chapter of Durham.

Figure 1.17. Painted Roll of Arms, from Matthew Paris's *Liber Additamentorum*; London, B.L. MS Cotton Nero D.1, fol. 171v; c.1244–55; reproduced by permission of the British Library.

Figure 1.18. The Royal Arms embroidered on a seal-bag of the City of London, with dragon supporters; obverse; c.1319; reproduced by permission of the Corporation of London Records Office.

Fig. 1.19. The Painted Chamber: Soldier-Guardians of Solomon's Bed, c.1263–67; watercolor copy by Edward Crocker, c.1819; see fig. 1.3 for positioning; reproduced by permission of the Ashmolean Museum, Oxford.

Figure 1.20. The Wilton Diptych, outside panels; c.1395–1413; note crowned lion and heraldic shield bearing the arms of Edward the Confessor impaled with those of England and France, quartered; also note the white hart of Richard II, the most notorious retaining badge of the period; reproduced by permission of the National Gallery, London.

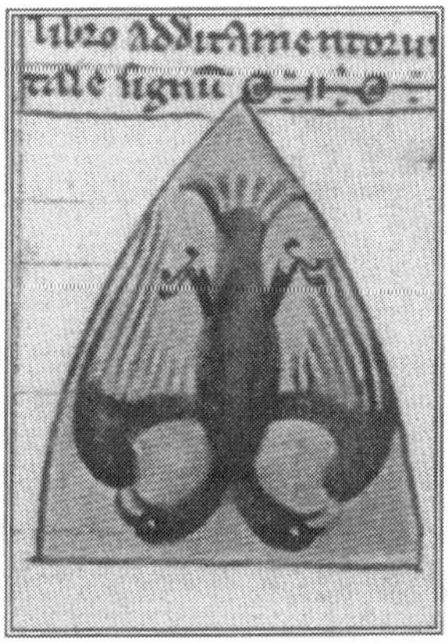

Figure 1.21. Inverted heraldic shield of Emperor Frederick II (to indicate his death), from Matthew Paris; Cambridge, Corpus Christi College MS 16, fol. 245v; c.1250; note imperial eagle as on Richard of Cornwall's 1257 arms, fig. 1.11; reproduced by permission of the Master and Fellows of Corpus Christi College, Cambridge.

Figure 1.22. Christ in Majesty with Symbols of the Four Evangelists; Cleveland Museum of Art, Mr. And Mrs. William H. Marlatt Fund 1945.132; miniature excised from a London Psalter, c.1250–60 (per N. J. Morgan); alternately, the Abbey of Peterborough (?), c.1325; ink, tempura and gold on vellum; note resemblance to recto Psalter Map head of Christ, fig. 1.7; reproduced by permission of the Cleveland Museum of Art.

Figure 1.23. Psalm 109, Christ Seated Blessing/The Trinity; Cambridge, Emmanuel College MS 116 (2.1.6), fol. 191v; Oxford (?), c.1260–70; reproduced by permission of Emmanuel College..

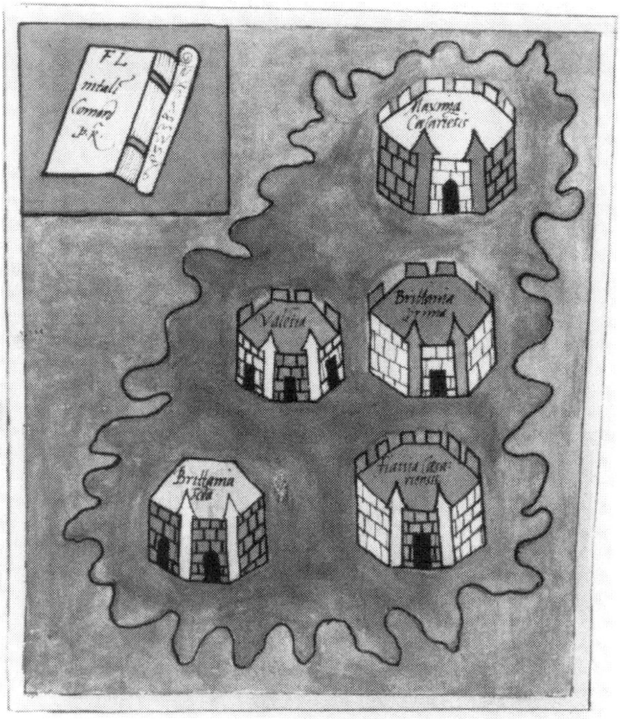

Figure 1.24. The *Notitia Dignitatum*, Britain; Munich, Bayerische Staatsbibliothek, Clm. 10291, fol. 212r; sixteenth-century copy (at several removes) of fourth-century original; note lack of concern for technical cartographic precision, seen in schematic coastlines and incorrect arrangement of the five provinces (Maxima Caesariensis, whose capital was London, should be in the southeast, not the northeast); reproduced by permission of the Bayerische Staatsbibliothek.

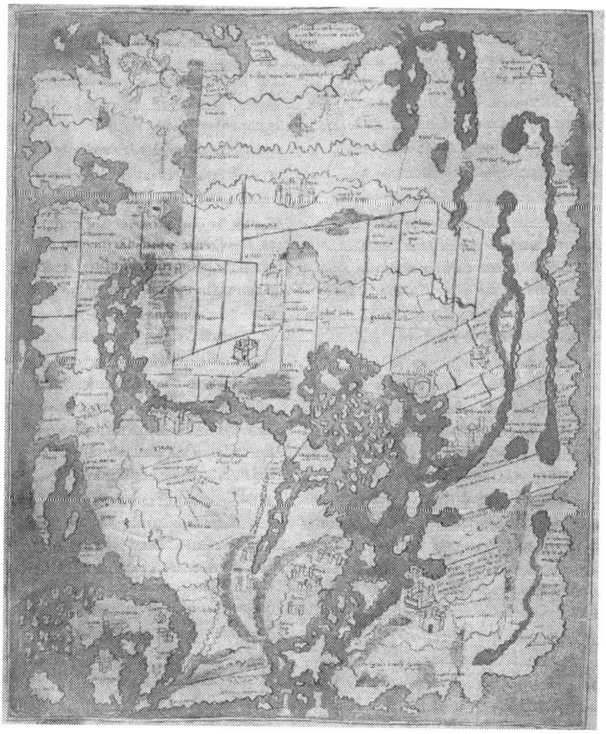

Figure 1.25. Cotton/Anglo-Saxon Map; London, B.L. MS Cotton Tiberius B.V, fol. 56v; eleventh century; note rectangular shape and prominence of Roman provincial boundary gridding; reproduced by permission of the British Library.

Figure 1.26. Genealogical Line of English Kings (William I to Stephen), each bearing an architectural detail, from Matthew Paris's *Historia Anglorum*; London, B.L. MS Royal 14.c.vii, fol. 8v; St. Albans, c.1250–55; compare town-icons to those on the London-Apulia Itinerary, figs. 1.4–1.5 above and 2.5 below; reproduced by permission of the British Library.

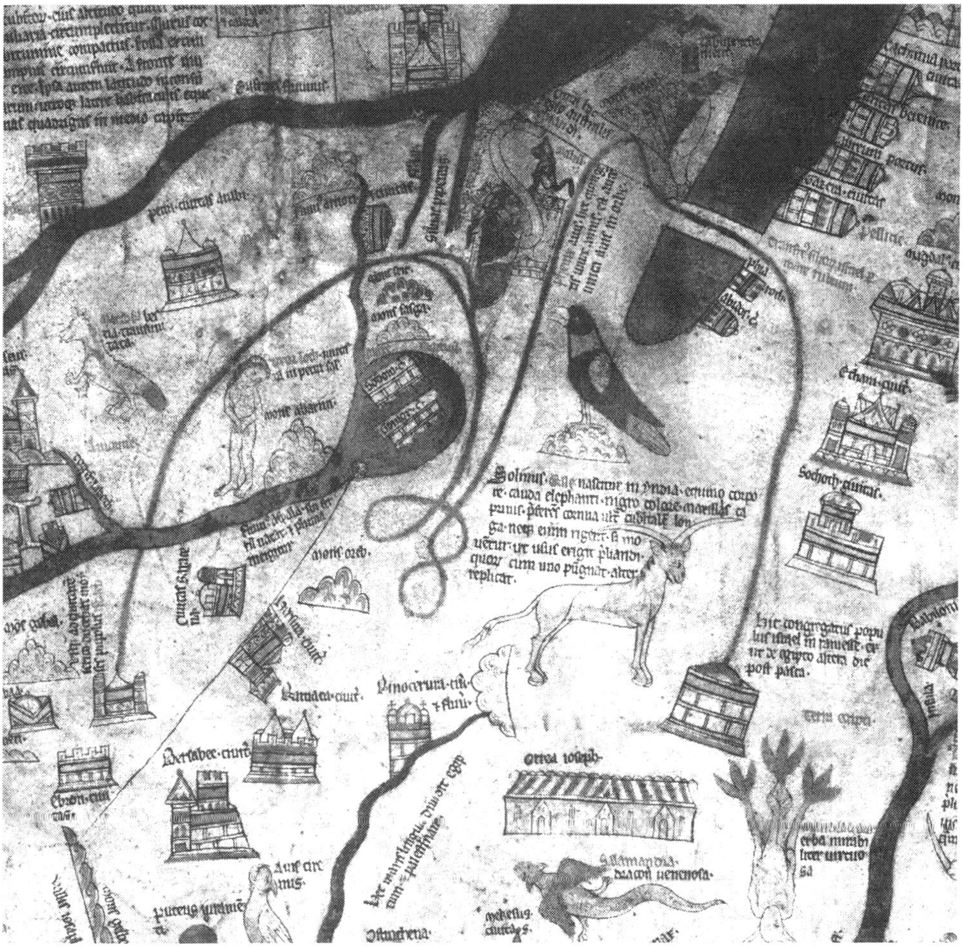

Figure 1.27. *Mappaemundi* as "at least as much written as drawn," with text as pendant or caption to icon; detail from the Hereford Map (Asia Minor); note the line marking the route of the Exodus, looping to indicate wandering in the desert; reproduced by permission of the Dean and Chapter of Hereford and the Hereford Mappa Mundi Trust.

Chapter Two

Figuring Britain

Regnal Maps and Matthew Paris

It is "a commonplace of medieval and social theory," as Sarah Beckwith has observed, "that the body is the image *par excellence* of human society." Beckwith finds Christ's body, in particular, "a key focus for the relationship between identity and the social structure for which it so often serves as an emblem." We have seen that a mid-thirteenth-century London bookmaker found a pair of Christological disc-maps, drawn by different hands and occupying two sides of the same leaf, to constitute (along with a set of prefatory illustrations later added) an appropriate Psalter frontispiece. But the politically charged architectural context to which these illustrations arguably bear witness—Westminster's 1230s–1260s Painted Chamber—suggests that if Christ's body functions, as Beckwith contends, less as an unproblematic image of unity than as "an image of social ordering" which "entails and occasions debate about the relation of sacred and social power," so may it be, similarly, that the body of the king functions, especially in the context of what may be called the quasi-heraldic realm of royal cartographic iconography.[1] Much like Christ's body, the king's body may serve, in other words, as a site for the contestation of political ideologies (primarily baronial versus royalist but also ecclesiastical versus imperial), especially as they relate to territoriality and to developments in administrative practice and state formation or definition.

Theological notions were employed to define the state, and to articulate various aspects of its nature, throughout the Middle Ages. What the thirteenth century in particular saw was the adaptation by the secular state of the ecclesiastical concept of the church as *corpus mysticum*, as the "mystical body" of Christ. In the view of Ernst Kantorowicz, efforts in the second half of the century "to hallow the secular polities as well as their administrative institutions" (in extension of Frederick II's initiatives in Sicily and the Empire) came increasingly to be based in the practice of "setting the state as a 'body' over against the church as a body": "the church as the supra-individual collective body of Christ," that is, "found its exact parallel in the state as the supra-individual collective body of the prince."[2]

In thirteenth-century England, *mappaemundi* in comparative abundance were available to represent *ecclesia*, to graphically figure the concept of the world church as the supra-individual collective body of Christ. Prior, however, to a series of conceptual breakthroughs made by English cartographers in the middle and late century, in which the island of Britain begins to emerge, comparably, as a body complete in itself—indeed one having something like a *mappamundi*'s (hence universal *ecclesia*'s) graphic proportions—no graphic rendering of an English *regnum* or kingdom as independent and abstractly mappable territory yet existed. It will be the business of the following pages to identify some of the links between English cartography, English kingship, and English territoriality in the second half of the century, years in which Edward I was consolidating his administrative hold on the land he inherited through increasing legal, fiscal, and bureaucratic centralization, even as he undertook a program of relentless imperial expansion in the British Isles.

KINGSHIP, PART THREE: THE FISC

We have focused our attention so far on the genre most prevailing in the cartographic imagination of the Middle Ages: i.e., the *mappamundi*, that ecclesiastically framed visual and geographic encyclopedia which has likewise proven so compelling to medievalist scholarship (largely on grounds of its departures from the modes and conventions of modern mapping). As we have seen, a number of prominent thirteenth-century world maps possess direct connection to England's royal house, indeed to Henry III and Edward I themselves. But at least as relevant, for the terrain we are entering, is that second major order of premodern terrestrial map, the map of the realm (*mapparegni* or "regnal map," in my usage) whose preeminent early exemplar is found in the Gough "Road Map" of Britain. This cartographic form would take on increasing political significance as England moved into and beyond Tudor times. What may be most interesting about this most celebrated medieval version of the regnal map, for our purposes, is that here again we have a prominent map that connects to the English Crown—this time not by way of architectural and artistic patronage, but instead via that primary bureaucratic institution through which a medieval king articulated his influence: his body of royal clerks. The evidence tying the Gough Map to royal sponsorship and interest, while compelling overall, is less direct than that which establishes royal ownership of *mappaemundi*, in the sense that there survives no documentary smoking gun, as for Henry III's mural or Edward I's *pannus depictus*. However, given the insular (as opposed to global or universal) scale and focus of this map, and considering as well the administrative and secular-symbolic nature of its geographical and visual content, we shall see that the Gough Map—highly flexible as a cartographical document, and astoundingly fertile as a political text—reaches well beyond the capability even of King Henry's mural *mappamundi*, in terms of its ability to advance arguments that intertwine topography and English royal authority.

The single medieval copy of this map now extant—that is, the actual document known as the Gough Map—has been dated to c.1360, on paleographical, orthographical, and visual stylistic grounds, as well as on the internal evidence provided by a handful of geographical features (e.g., place-names unknown before mid-century) whose inclusion establishes a definite *terminus pro quem*, for this particular incarnation of the image at least. Of course, however reliable, such a dating in no way invalidates (or need conflict with) the notion—espoused by a number of authorities over the years—that the map's basic graphic contours and underlying geographical data were probably compiled and set in form much earlier. According to these suggestions (and as arguments found later in this chapter and in chapter three will develop) Gough's prototype should be placed in the reign of Edward I, c.1280, with data corrections/accretions and visual embellishment being incorporated from the moment of this initial compilation intermittently forward.[3] Given that I have been obliged to work from photographs and facsimile reproductions, my contributions are not designed (nor will they be taken) as the first or last word in terms of the Gough Map's provenance and materiality. Instead I look to elaborate upon and give cultural and historical context—also ideological weight and implication—to the dating propositions and thematic observations of my forbears in study of this document. In any case, what is crucial to establish here is that the Gough Map's effective "date of composition" stretches across some eighty years, from a prototype developed in the reign of Edward I to an extant copy dating to the time of his grandson Edward III. Finally, because the extant version of Gough displays significant local knowledge in the vicinity of Lincolnshire and South Yorkshire, and because later maps derived from it (or a common exemplar) contain specialist knowledge of other regions, historians of English cartography have frequently posited that multiple versions of such a "national" map may have existed simultaneously in medieval times, designed for and held at various regional centers.

E. J. S. Parsons judged in 1958 that "there is no doubt that its purpose was to serve as a map for travellers," and no commentator since has dissented (to my knowledge) from characterization of the Gough Map directly along these lines. Going further, map historians have supposed that between "symbolic" *mappamundi* (like Edward I's *pannus* at Westminster) and "practical" *mappaeregni* (like the Gough "road map") nothing less than a "cartographic revolution" must have taken place—never mind such maps' contemporaneity and apparently shared patronal contexts. Medieval England's postulated cartographic revolution, as it is generally understood, unfolds silently and almost entirely off-stage. Gough, the harbinger of modern mapping, springs fully formed, as it were, from the head of some king's clerk with a genius for cartography (and a ready fund of geographical data), with little or no precedent to announce it formally and with no concomitant epistemological shift to place it ideologically and explain its appearance here as opposed to any time or place else.[4]

A turn to the examination of some formal cartographic genealogy should begin to make contemporaries of the king's two maps—or rather map-types, *mappaemundi* and *mappaeregni*—and to underline these genres' social comparability and complementarity. Before this, however, and before we turn to the Gough Map in earnest in chapter three, further consideration is due to kingship—specifically, to a new "duality" within the concept whose appearance in England has been located to the second half of the thirteenth century. This new incarnation of a royal *geminatio* differs from earlier kingship's twinned sacred and secular bodies in that it derives not from the office's "mixed" nature (i.e., its dual temporal and spiritual aspects) but rather from the establishment of what Kantorowicz calls an "extra-terrestrial or extra-feudal realm within the realm," a kind of eminent public domain of rights and properties which pertained to all but lay under the special protection of the king.

This "supra-personal" royal public sphere was known to medieval legal and political writers as "the fisc," a term which derives ultimately from Roman law. In the neat formulation of Gratian (375–383), the terms *Christus* and *fiscus* are linked homophonically even as they are opposed conceptually, in the emperor's closing epigraph on the subject of that due to be rendered God and Caesar: *Hoc tollit fiscus, quod non accipit Christus* ("What is not received by *Christus*, is exacted by the *fiscus*"). This usage pairing *Christus* and *fiscus* reappears frequently in medieval legal texts.[5]

In the Carolingian period "the fisc" was a common term which denoted simply the king's personal, private property. This definition (fisc as royal purse) lingered to some extent, but over time the concept accrued associations more impersonal and public. On most points consensus was elusive. For instance, the fisc was considered by some jurists to be "detached from the person of the ruler" and thus to comprise a kind of corporate body which had (at least potentially) an independent existence itself. But the fisc was also understood to be represented and administered (if not indeed owned) by the king, and to be in some matters coincident with him, indeed an embodiment of his sovereignty. That is, the fisc had its own contours, but this body was somehow and at once also a royal body.[6]

On one hand, the fisc was comprised of abstract qualities such as peace and justice, components of the public good made territorially manifest, we might say, by the running along principal highways of the king's peace, or in the need for Welsh litigants to come to London (once Wales had been overrun 1277–83) to conduct their legal business. But the fisc also included more concrete "public things": for example, "utilities such as public roads, public waters, or public squares." By the mid-thirteenth century, in any case, the concept of an "independent and impersonal fisc," a kind of abstract royal body coincident with the rights and community of the realm, had developed. On the model of church property, fiscal property was inalienable, and the fisc itself was perpetual and ubiquitous. Things fiscal, because touching all, were "things quasi-sacred"—they were holy on their own terms: social and political, not ecclesiastical.[7]

What Kantorowicz ultimately provides, for the English thirteenth century but with transitional reference to Frederick II's Holy Roman Empire, is the model of a basic shift in conceptions of kingship. The Christ-centered, sacramental earlier conception, in which the king "becomes deified in consecration," gives way to a law-centered, fiscal later conception, in which the king becomes the "supra-individual administrator of a public sphere." From *vicarius christi* ("vicar of Christ"), the medieval king became *vicarius fisci* ("vicar of the fisc").[8]

Such developments in legal and political theory have relevance to contemporary mapmaking because in order to achieve the crucial "hallowing of its essence" which would set it apart from the church (spiritually and institutionally), the sovereign state needed to become "bodified" itself. And what the development of a legal theory of the fisc enabled ultimately was a new understanding (and with it a new visual expression) of the concept of *patria* (a "fatherland" or civic "home" to which allegiance is due) as territorial rather than urban. That is, *patria*'s classical basis in the concept of the city (Rome as the seat of empire) came in the course of the thirteenth century to yield, according to Kantorowicz, to a concept of *patria* based in the kingdom or monarchy—the defense or protection of which came to have "semi-religious connotations comparable to the defense and protection of the sacred soil of the Holy Land itself."[9]

How English maps reflect and extend the terms of this shift in conceptions of kingship and territoriality can best be seen in the analysis of formal overlap between map-types, as it is here that such political ideas are, in the process of their working out, given incremental graphic expression.

MAP GENRE OVERLAP: FORM

In 1978 Juergen Schulz argued for a division of early maps into "broadly speaking" two categories: on one hand those "of a narrowly cartographic content, the function of which must have been simply to report geographical and topographical facts"; and on the other those "with an ideal content, material that must have had a didactic intent." If word-choice has varied over the years, the tendency to invoke such a terminological binary opposition remains: where

Schulz spoke in terms of "mystical" as well as ideal and "technical" as well as narrowly cartographic, the semantic register of Harley and Woodward (1987) includes, for example, "paradigmatic" in opposition to "factual." *The History of Cartography*'s landmark status in the field and role as a guide for research makes its terminological choices—and the interpretive framework these pairings reveal—especially noteworthy. Much favored here is the generic "practical," although more careful terms like "geographical way-finding" are sometimes substituted; while "symbolic" alternates with and at times threatens to collapse into the "religious" or "cosmological," against which—disconcertingly—a "secular" map's necessarily "practical" nature is frequently set (as if secular and symbolic, or religious and practical, were somehow terms at odds).[10]

The arguable complementarity of the various genres of medieval maps in terms of social function—particularly with reference to the discourse of crusade—has been explored above, in chapter one ("A Crusader Atlas, or, Map Genre Overlap: Function"). Simply in itself, this functional overlap begins to break down the practical/symbolic split perceived to be constitutive of the nature of medieval maps uniformly by type. But even more pronounced than their categorization into those symbolic and those practical have been critical efforts to keep the various genres of medieval maps firmly isolated from one another also on a strictly formal level. In the face of considerable evidence of hybridity—a quality more in evidence at some times and places than others, certainly—critics have still largely clung to the neat notion that overall medieval maps fall into "several quite distinct traditions," "mainly discrete groups" with only "seldom [any] contact between them."[11]

A deep pinprick marks the very center of the Hereford *mappamundi*. Whoever laid out the map apparently used this point to ground one leg of a drawing compass, whose other leg was employed to draft both a small, inner circle—representing the walls of Jerusalem (fig. 2.1)—as well as the several lines marking ocean and border at the outer circle of the map (see fig. P.3, above). If historians of cartography have judged formal (like functional) overlap between *mappaemundi* and other medieval map genres to be mostly negligible, then the pinprick at the center of the Hereford Map, and these inner and outer concentric circles making similar bodies (similar in the geometric sense) of the Holy City of Jerusalem and of the sacramental incarnate world, together begin to insist otherwise. They serve notice, on a basic geometric level, of substantial borrowing between medieval cartographic genres—recognition of which is crucial to the social interpretation of late-thirteenth-century English maps, the Gough Map in particular.

After world maps and portolan charts, maps or plans of Jerusalem (*sitii Hierusalem*) comprise medieval Europe's most numerous cartographic documents. In basic form most of these plans do not differ significantly from the

Hereford Map's iconic rendering of Jerusalem as a walled, circular town crested with battlements and crowded with gates and buildings—except that fuller plans render the main thoroughfares of the city in a cross-set scheme, which lends the composition further visual structure and divides the city into three or (more often) four sectors (see fig. 2.2). As even a glance makes clear, in this arrangement the medieval map of Jerusalem makes a spatial argument very similar to that made by the (cruciform) medieval *mappamundi*. Just as the inner circle of Jerusalem sits atop the world map's T-bridge convergence of waterways (figs. P.3 and 2.1), an inner circle demarcating the *Templum Domini* or Temple of the Lord typically rests above the central intersection of the Jerusalem-plan's two main avenues—one running horizontally between the Damascus Gate and the Gate of Mount Sion, the other rising vertically into this from the Tower of David at map's bottom (figs. 2.3 and 2.4). For that matter, some maps of the Holy City even have, after Hereford's fashion, an unmistakable pinprick or compass-grounding at their centerpoint (see for example the Brussels Map of Jerusalem, Bibliothèque Royal Albert MS 9823–24, fol. 157).[12]

This overlap in simple geometric structure between *mappaemundi* and *sitii Hierusalem*—the latter are set in crusader manuscripts even more often than the former, incidentally—suggests that a single formal concept of what a map was, or should be, informed at least two of medieval cartography's purportedly "distinct" generic traditions.[13] From the above evidence it would seem that a map has, ideally, three geometric qualities: it is circular (although further encased by an orthogonal frame or manuscript page); it is centered (upon an isolated point, an inner circle, or an intersection of routes); and it is radial (whether these spokes represent rivers, actual roads, or simply overland town-list thoroughfares).

On this final point, it should be noted that a medieval map's spokes, to the extent that they are cruciform, are lines that have been schematized. Importantly, however, these lines or town-lists also represent actual travel routes to the map's center or spatial hub, and as such are not just symbolic, ideal, or paradigmatic but also practical and factual, the physical trace of wayfaring. In addition to an ideological weight—these are maps of exhortation after all—world- and Jerusalem-map radii bear the inscription of (as well as set the course for) military, mercantile, administrative, pilgrim, and other traffic.[14]

Of course, not all medieval maps, not even all medieval terrestrial maps, have this form. Itinerary maps constitute another basic group. Maps of this type, as we saw via Matthew Paris in chapter one, typically chart a single route or journey (e.g., London-Apulia), often with no regard for the concept of direction (whether relative or absolute), and often in a series of strips or manuscript pages. Essentially these maps are elaborations, in greatly varying degree, of classical or medieval non-figural town-lists. Sometimes very little in the way of detail is added (topographic or symbolic), sometimes a great deal. More important is that on some maps, formal innovation is evident such that the basic itinerary (or constitutive sequence of towns) serves not simply to mark or delineate a route but as the backbone of a more fully annotated geographical document—what may be called an incipient territorial map (i.e., one with an inclination toward representation of multi-dimensionality).[15]

The divergent epistemological character of maps which occupy different positions along what may be envisioned as an itinerary-to-territory cartographic continuum—even if these are not positions so terribly far apart—can be seen clearly in a comparison of Matthew Paris's London-Apulia Itinerary to his Map of Britain, both of which are extant in four versions. These maps are contemporaneous; both Itinerary and Britain (in all versions) date to c.1250–55, which is likewise the period of Paris's (lost) sketchbook copy of the king's Westminster *mappamundi* and of his reference to two other English world maps. Yet the latter of this pair, Paris's Britain, is formally a far riskier undertaking than the other, and overall a much more innovative cartographic proposition—"proposition" because, as social geographer David Harvey put it years ago, a map is in one sense just a theory about the world of social relations.[16]

The first of our pair of maps, London-Apulia (figs. 2.5 and 1.5, above), includes some non-essential information, such as an architectural sketch to mark each site noted plus an occasional touch like the stork nesting in the towers of Susa (fig. 1.4, above). Still, this production may be characterized as a classic itinerary map in that it provides only such information as may be of interest to the actual, prospective, or ideological traveler along this route, while eschewing that which lies off it. Frames encase each strip but there are few topographical borders—only a line segmented by buildings.[17]

Paris's Maps of Britain, on the other hand, while clearly constructed on the basis of a single route—it runs from Newcastle at top down to Dover at bottom, and like a proper itinerary does not deign to chart a necessary turn eastward at London (see fig. 2.6)—yet display much interest in topographical and settlement features which lie off this main axis. Indeed, to differing degrees in each version (see figs. 2.7 and 2.8) the Map of Britain begins to place its geographical icons in relative spatial (not just sequential) proximity to one another: we stand on a plane, not a line, and a concept like triangulation begins to be possible.

Although Paris comments famously in the margin to one version that his map's shape would look differently had the dimensions of the page allowed,[18] the fact that there is any attempt made at all to encase this route territorially, to sketch the shape of the island as a whole, must be seen as an extremely significant structural feature. This attention to coastal outline comes at the expense, formally speaking, of the severe linearity of the standard itinerary map, some of whose myopic scope and function as a prescription to action falls away.

Such a map's mandate, in short, is to do far more than simply chart the route (in a "practical," wayfaring sense) from Newcastle to Dover. It has also, or instead, the ambition to describe Britain as a whole island, as an expanse of (political) territory.

But if high medieval terrestrial cartography's two basic layouts were the linear single-itinerary map and the circular, centered, and radial territorial map (representative of world or city), Matthew Paris's attempt at a regnal or provincial map (territory on an intermediate level, part-way between *mundus* and *urbs*) is no longer one and not yet quite the other. The linear basis of its single-route construction is at cross-purposes with its ambition to chart multidimensional territory, a goal seen clearly in the island's breadth, extra-axial features, and coastal outline. Formally, the map must be regarded as something of a failed compromise—hence, perhaps, the tinkering of four versions, each of which annotates the axial route more fully. Even so, the recurrent attempt to sketch multidimensional regnal space, an island with breadth where multiple itineraries are possible, is on each map in the set decisively undercut by insistence on construction along the one axis only, Newcastle to Dover. Even with the help of coastal features, the view from a single route can only imperfectly imply expanse.[19]

Michel de Certeau has characterized the itinerary and the map in terms of a "bi-polar distinction," and in Paris's Maps of Britain we have seen a failure to resolve the tension between such poles. However, not all medieval attempts to reconcile these polar cartographic modes result in such a textualizing of formal dissonance. Plans of Jerusalem, as we have seen, use local streets to help give body or structure to the sacred space of the Holy City, in urban-planning extension as it were of the cruciform church. And often *mappaemundi* go far beyond a simple T-O schematization of routes to world and map's center. Rather, as Gerald Crone has demonstrated through examination of the Hereford Map's toponymy—cities are a dominant geographic feature especially in Europe, he notes—medieval world maps were, like their antique prototypes, apparently constructed directly on the basis of a series of originally non-figural (i.e., written) classical, biblical, legendary, and contemporary medieval itineraries. Lists of towns were lifted, that is, from administrative texts like the Antonine Itinerary (used also in the *Notitia Dignitatum*); from the ministry of the Apostle Paul in Asia Minor; from accounts of the travels of Alexander the Great in the east; and from contemporary medieval routes, mercantile (e.g., the movement of English wool in Gascony), military (Edward I's progression through Europe on crusade 1270–73), and religious (pilgrimage itineraries to Rome, Jerusalem, and Compostella). Plotted onto the outline provided by the traditional T-O schemata, these routes, in their plurality, serve not as an alternate geometric ordering feature which deflects attention away from the prevailing concentric visual argument of the image, in the way that backbone itinerary and coastal outline compete for compositional primacy, as it were, on Matthew Paris's island. Rather, in comprising the structural material from which are fashioned the spokes or radii linking the composition's center with its outer circle (and in so doing, establishing a kind of grid pattern analogous to Rome's outward-churning but inward-returning road-network and land-centuriation system), these routes contribute to the world map's simultaneously centrifugal and centripetal imperial mode.[20]

This point of the importance of a plurality of routes in the representation of territoriality may be best illustrated by the *Tabula Peutingaria* or Peutinger Table, the medieval copy or version (c.1230) of a Roman map designed to provide "a view of the world" from apparently "about the time of Augustus."[21] Like Matthew Paris's (nearly contemporary) Maps of Britain, the *Tabula Peutingaria* betrays a decided tension at the competing formal demands of the linear itinerary map and the radial, centered, and circular territorial map—although such tension plays in this map a constitutive rather than a deconstructive or destabilizing role. For unlike Paris's neither one thing nor the other attempt at a regnal map, the *Tabula Peutingaria* strikes a formal compromise which allows it to succeed both as a representation of expansive territory and as a register of discrete itineraries. In layout this "map" is composite, not a single coterminous image but a serial roll of long, horizontal strips charting routes in the Roman provinces (fig. 2.9).[22] The *Tabula* achieves a sense of unified territoriality, however, simply by invoking the standard T-O image of medieval cartography. What is crucial is that it suggests circularity without having to carry this long-entrenched spatial mode to its usual formal conclusion (i.e., it need not draw a circular outline itself, only allude to it).

The section of the map containing Rome does most of this work (for a close-up see fig. P.4, above). In a detail which recalls province-insignia from the *Notitia Dignitatum*, the city is represented as a personification in full regalia, orb in hand and set enthroned within a circular shell, a hub from which the empire's twelve main routes radiate, so many spokes in suggestion, as it were, of a wheel-shaped world. The *Tabula*'s circular Rome recalls medieval Jerusalem's world-map and city-plan circularity, of course. And just as is the Holy City's, the ancient Seat of Empire's geometry is typological: implied is the concentric circularity of the world, precisely as medieval *mappaemundi* and images of the Host present it (see also fig. P.7, where the creator uses a compass to sketch a concentric cosmos).

At a typological level, then, the *Tabula Peutingaria* cites the circular, centered, and radial map-icon, familiar from *mappaemundi* and *sitii Hierusalem*, as the geometric form which is needed to make sense of its serial text, to bring a defining imaginative order to its set of discrete and linear provincial itineraries. The *Tabula* is like a *mappamundi* that has been disassembled; if presented spoke by spoke, it is still to be understood, spatially, in terms of a wheel or disc.

Why does this matter to conceptions of English kingship? It matters because the graphic concept which this map grasps while Matthew Paris's four Maps of Britain do not, is that multidimensional territory—whether *mundus*, *imperium*, *urbs*, or *regnum*—can, if based in linear, town-list construction, be sketched successfully only by way of multiple, collated itineraries. A single backbone will in most cases be insufficient, for such a map remains largely a journey even if and as it seeks to imply a body. Recall the camel pointing the way not on circuit around but *through* Matthew Paris's Palestine toward centrally placed Jerusalem, and how this map formerly was read as an extension of the London-Apulia Itinerary (see fig. 1.6, above). Paris's Britain, like his Palestine—they appear on consecutive folios in one manuscript—is thus terrain ultimately to be moved across, if in the meantime lived within. This is territory, moreover, defined inescapably by its relation to a theological, internationalist spatiality, by its citation or acknowledgment of the ruling geographies of Christendom: those exhorting crusade, those presenting a universal *ecclesia* (or alternately *imperium*), to which metropolitan body's interests the regnal or regional territory is necessarily subordinate, and against which it stands as no independent body itself, only a body part or single provincial itinerary.

That Matthew Paris's Maps of Britain should falter in their attempt to reconcile, for territory on a regnal level, the competing formal demands (and divergent ideological implications) of the linear itinerary and the circular/centered/radial map, may highlight the degree to which there flourished a defining tension between ecclesiastical claims to universal sovereignty and secular state articulation in England in the mid-thirteenth century. Sometime in the last few years of his life, however, Matthew Paris sketched one further cartographic image of Britain, one which has only rarely received notice, perhaps in part because it appears uniquely near the end (f.187v) of a commonplace book and portfolio he kept from c.1247–59 (compiled beginning c.1251–53 into the *Liber Additamentorum*, B.L. Cotton Nero D.1) rather than, as for example three of his Maps of Britain do, in a deluxe copy of the *Chronica Majora* or one of its abbreviations.[23] This work is only a draft (a spare one, at that) and has none of the pictorial elaboration or ripe commentary for which Paris is famous, although it seems to me possible that fuller versions have been lost, given the precedent of multiple versions of his other maps (such as an abbreviated London-Apulia Itinerary in this same commonplace book, ff.183v-184).

But the *Scema Britannie*, a "Diagram" or "Figure of Britain," may yet be Paris's most innovative piece of cartographic work. For this sketch textualizes precisely that conceptual breakthrough which appears to be responsible for, or at least to structurally underlie, the subsequent advances in terrestrial cartography which would soon prove able to represent Britain as a full territorial unit itself (i.e., no longer as just an itinerary or part of a larger body), a state achieved fully in the Gough Map and becoming afterwards (certainly by Tudor times) a standard political image. Indeed, if the world map has already in Henry III's time taken on a heraldic association, the regnal map becomes in periods not long thereafter monarchy's special insignia, arguably the state's defining image.

The c.1255–59 *Scema Britannie*, we might say, corrects the formal shortcomings of Matthew Paris's earlier Maps of Britain. It builds in part on the multiple and convergent road-radii spatiality of the Peutinger Table (which is, recall, recopied by a scribe of the previous half-generation or so); and in part on the ideological and geometric precedent of contemporary, quasi-heraldic royal world maps such as the Westminster *mappamundi* whose presence Paris records. And in mobilizing such an inheritance, the *Scema Britannie* seems to set the stage for the stunning departure from cartographic convention that comes in the Gough Map of Britain. A new kind of cartography crystallizes in Gough, a map whose prototype was to appear less than a generation after Matthew Paris. But if the technical and conceptual cartographic development prerequisite to the Gough Map has never been adequately set out, this may be because the *Scema Britannie* has received inadequate (often dismissive) critical attention.[24]

The *Scema* is neither a visually complex nor a densely encrusted image, as figure 2.10 shows readily. Its construction has in essence three major components—and not coincidentally, we have seen these very geometric features likewise define both contemporary *mappaemundi* and *sitii Hierusalem*. Paris's half-page commonplace-book sketch of Britain, first of all, is centered—at or near Dunstable, a site highlighted or set-off in a rectangular frame as are only two other names on the map (*Totnes* in the extreme upper-left corner and *Salesburia* in the far west, near the map's top). Second, the actual matter of the *Scema*'s geography consists entirely of four criss-crossing itineraries (labeled, described, and book-ended in some cases by place-names) which run from coast to coast in clean geometric division (vertical, horizontal, and two diagonals) and which intersect, if not directly in a point upon Dunstable, at least in its immediate vicinity.

Third and finally—this point is less obvious than the others—the composition as a whole is given a kind of in-curving schematic border. If not quite a circle, the map is an oval, with tails in the northeast (bottom right) and southwest (upper left) added either as a small measure of (attempted) empiricism or more likely to make room for the text that is crowded more thickly here (along Fosse Way) than anywhere else on the map.[25] It may be, as this is the route north through Lincoln and hence a portion of that very itinerary which Paris employs as a backbone for his other maps of Britain, that he had more information for this than he did for other routes, or that such information came more readily to mind in what appears to be an offhand sketch (i.e., one not compiled through sustained research; the sketch is never "filled in" geographically or pictorially). In support of this reading it may be said that the Figure's outline—its coast or regnal border—appears to have come only secondarily in the composition process, after the first inscription of the routes; not vice versa or more or less simultaneously, as seems to have been the case on the Maps of Britain, whose coastal details can be quite elaborate (they are jagged and empirical more than schematic or geometric) and which have apparently been taken from corresponding sections of one or more world maps.[26]

Matthew Paris's description of his work as a *scema* of Britain—the phrase is part of a curving inscription above the figure's western or top edge—provides more than just a convenient title. It also gives indication of this work's ultimate textual source, and may thus help orient interpretation. On one hand, this rubric sets out yet again, if read in its entirety, the proposition that a single north-south axis may serve as the defining basis or primary backbone of Britain's geography: *Nota quale est Scema Britannie cuius mete finales sunt tothenes quae est in cornubia et Cathenes in Scotia* ("Such a mark as this is a Figure of Britain, whose measured ends are Totnes in Cornwall and Caithness in Scotland").[27] More important, however, is that this *scema* or "diagram" which the page below goes on to delineate may also be understood according to the primary, rhetorical meaning of the term. Matthew's "mark" or *nota*, that is, may be taken as a graphically rendered *scema* in its sense as a "figure of speech."

The four structurally defining routes in Matthew's figure, this is to say, are not simply random thoroughfares for which he has found town-list data or textual precedent individually. Nor are they contemporary routes the layout of which he has gathered sufficient empirical evidence to support (as seems to be the case on his Maps of Britain, for which his "principal sources . . . were doubtless the reports of travellers").[28] Instead, the four routes which comprise the *Scema Britannie*'s eight spokes are in fact the component parts of a common "figure of speech" themselves, the conventional elements of a passage standard in medieval chronicle-writing in Britain from the early twelfth century onward. Typically placed in the account of the reign of lawgiver and roadbuilder Belinus (who established a "kingship of the entire island from sea to sea" and then went on to conquer Rome), the matter in this passage is imputed by Geoffrey of Monmouth to have appeared in King Alfred's translation into English of the "Molmutine" Laws (named for Belinus's father Dunvallo Molmutius) which the Romano-British historian Gildas translated from Welsh into Latin.[29] Beginning with Henry of Huntingdon's first version of his *Historia Anglorum* (c.1129–33), the trope of the Four Ways appears regularly in both Latin and vernacular histories, including Geoffrey of Monmouth (c.1136), Robert of Gloucester (d.1147), Wace (c.1155), Layamon (c.1199–1225), Roger of Hoveden (c.1180), Walter of Coventry (c.1293), Peter Langtoft (d.1307), the *Short Metrical Chronicle* (c.1315), and Robert Mannyng of Bourne (c.1338), all the way through to Ranulph Higden (c.1344) and beyond.[30] Significantly, the account of King Belinus's ancient road-building is not present in Roger of Wendover's *Flores Historiarum* (c.1204–34), the St. Albans universal chronicle to which Matthew Paris's *Chronica Majora* serves as an adaptation and continuation for the years 1235–59. Yet Paris describes himself explicitly as one "who [has] read the succession of historical annals," a claim which is borne out, for example, by an extant manuscript of Geoffrey of Monmouth's *Historia* which contains glosses in Matthew's own handwriting. Strong evidence, likewise, suggests Paris knew and used Henry of Huntingdon and in particular this cartographically pregnant material.[31]

The passage in question differs somewhat between versions but in essence rehearses the run of the four main highways of Britain in order to evoke the island-kingdom's expanse and as a means of establishing its ruler's coast-to-coast authority. The point is that the "whole realm of Britain" (so Geoffrey of Monmouth) is law-inscribed territory, comprised of "privileged" cities linked by new and conspicuously straight "consecrated" highways. But this realm is also, less auspiciously, the unhappy territory (*leode* in Layamon's key term, signifying both "land" and "nation") destined to be rent by intermittent civil war and, worse, overrun by the recurrent plagues of Roman, Saxon, Pict, Danish, and Norman invaders.[32] As Matthew labels them, his *nota*'s constitutive routes are *Ykenild strete* ("Hikenild Street"), running from Bury St. Edmund's to Salisbury; *Fossa* ("Fosse Way") from Totnes through Lincoln to Caithness; *Erningestrate* ("Ermine Street") north from Canterbury; and *Watlingestrate* ("Watling Street") northwest to Chester. Text-to-text variation notwithstanding—a given since in redaction the passage is alternately dilated and condensed—these routes are precisely those named or described in the bulk of the chronicles.

The *Scema*'s rendition of Britain as that figure of speech which treats King Belinus's program of road- and realm-building thus serves, implicitly, to link regnal territoriality to the chronicle tradition's twin key issues of invasion and conquest. English kings, whose reigns typically give structure to these narratives, receive blame or praise from native chroniclers in seemingly direct measure to their military/territorial failure or success. King John, for example, whose immense losses in 1204 included Normandy, Maine, Anjou, Touraine, and parts of Poitou, is uniformly bewailed as a "bad" king; in contrast King Edward I, a celebrated crusader who overruns Wales and nearly succeeds in annexing Scotland, tends to emerge as a "good" one. What is more, frequently in these texts the quasi-sacred land of the English kingdom cries out, when cleft by civil war or its borders penetrated, like a wounded sentient body itself. And when the land's governance is in turmoil, or its king dies or physically ails, the land of the kingdom likewise seems to ail, as for example with the passing of Arthur at the Battle of Camlaan, upon which plague and crop-failure ensue.[33]

To the extent that its structurally defining routes are the four highways familiar from the chronicle tradition's articulation of the lawful breadth of Britain, the *Scema Britannie* thus crucially reframes English territoriality, placing it within the context of a nativist discourse whose concern is for the reigns of kings and the fortunes of a British island-kingdom instead of for the relative position of a provincial Britain within the larger body of either universal *ecclesia* or Roman *imperium*. Yet so far unexplained is the strange fact of this map's western orientation. Note, within the squiggling line of waves around the island, the legends *Occident* at center-top, and *Orient* (or *Meridies*) at center-bottom, terms whose service as markers of abstract direction is underlined by the addition of their corresponding winds: e.g., *zephirus* (the west wind) above *Occident*.[34] No commentator has forwarded any suggestion to account for this key and unique formal detail. It may simply be Matthew Paris's idiosyncrasy at work, in a period of sometimes-uncertain cartographic convention; his other maps are oriented east and north variously. But perhaps this is another cartographic "choice" which should be put down to the constraints of a manuscript page, or more precisely to the spatial-codiographical conventions of medieval scribal and illuminatorial practice.

It is likely that our cartographer worked in this sketch, his *nota* giving a diagram of Britain, from his tabulated memory of the chronicle passage discussed above. As Mary Carruthers has explained, *notae* are marks or signs employed by medieval readers to translate textual information into visual terms, so that it may be fixed in an indexed memory. In manuscripts, *notae* tag passages of particular interest; but more importantly than as aids to browsing such signs function mnemonically, as keys to a spatial and systematically imprinted "memorial subject concordance."[35]

Given the survival of his glosses to a manuscript of the *Historia Regum Britanniae*, Matthew Paris might be expected to have graphed the highways of Britain in the order in which Geoffrey of Monmouth and most other accounts of Belinus's road-building (i.e., those following the *Historia*) typically list them: that is, beginning with Fosse Way ("which should bisect the island longitudinally from the Cornish sea to the shore of Caithness") and ending with Watling Street and Ermine Street ("two more roads in diagonal pattern across the island," as Geoffrey puts it laconically).[36] The habit of years of scribal activity may have led Paris to begin, following standard practice in Western Christian codiography, in the upper left-hand corner of his page. With a first line thus speculatively drawn (to conclude where a half-page of text might conclude, at the lower right of the sketch), the next route graphed appears to be set down with a more confident hand. *Ykenild Strete* is the straightest and shortest route on the map; it has (like *Fossa*) a place-name to ground each end; and it is segmented quite systematically with incremental text (i.e., equal characters each side of the line). As such, it looks to have been drawn prior to the two remaining routes, each of which has only a single place-name to bookend it (i.e., *Erningestrate* has no northern station, *Watlingestrate* none in the southeast) and a lesser amount of descriptive text, with what there is being distributed unevenly.

The sureness of pen characterizing the composition of *Ykenild Strete*'s course and text may have much to do with the direct correlation there will have been between graphing this route and drawing a standard itinerary map, along the lines of Matthew's likewise vertical and incrementally annotated London-Apulia Itinerary. But perhaps also contributing to this sketch's peculiar mode of composition (and resulting western orientation) may be the visual exemplar provided by medieval illuminated manuscripts in their arrangement of sequential-narrative scenes. The multiple-panel pages that are common in Psalter preliminary illumination-sequences, for example, proceed also in this direction when telling a story, beginning in the upper left and moving next across the top of the page.[37]

If Matthew Paris proceeded after this manner it may be that this map's western orientation chose him, as he moved left to right and eventually clockwise (following Geoffrey's Fosse Way to Watling Street/Ermine Street model), rather than that he chose it. Perhaps, that is, he began his first line with *Totnes* and his second with *Salesburia* (these are the only names encased in rectangles besides Dunstable at the center, the practice being abandoned thereafter) and drew his first route diagonally to where a block of text would end, and his second vertically as if it were a standard single-itinerary, with the third and fourth routes added in relation to these.

There are some problems with this account, however. Geoffrey of Monmouth does not give the course of the second road as Matthew here graphs it, but rather describes it "running west to east across the kingdom" from St. David's to Southampton (not coincidentally, a Roman road fitting just this description ran through Monmouthshire). Geoffrey himself does not give names, moreover, to any of his four roads (although some chroniclers following his version add them), nor does he describe the courses of Watling Street and Ermine Street even as roughly as Paris's sketch does.[38]

What is immediately clear from this is that while Matthew (indisputably) knew Geoffrey's *Historia* well, still his *Scema Britannie* clearly depended, whether primarily or secondarily, on a source beyond it. Almost certainly this will have been that source from which Geoffrey himself, although adapting the passage to fit his own purposes, "plainly derived" his material: that is, from his contemporary and apparent rival Henry of Huntingdon, whose work Paris also uses. Huntingdon's version deserves quotation at length:

> In such estimation was Britain held by its inhabitants that they made in it four roads from end to end, which were placed under the King's protection, to the intent that no one should dare to make an attack upon his enemy on these roads. The first passes from East to West and is called *Ichenild* [in some MSS *Ikenild*], the second runs from South to North and is called *Erninge Strete*, the third crosswise from Dover to Chester, i.e., from South-East to North-West, and is called *Watlinge Strete*; the fourth, the greatest of all, begins at *Totenes* and ends in *Catnes*, in other words runs from the commencement of Cornwall to the limits of Scotland, and this road passes across the island from the South-West to the North-East. It is called *Fossa*, and passes through Lincoln. These are the four great Roads of England, spacious in their dimensions, and admirable for their construction, protected alike by the edicts of our Kings and the written laws of the land.[39]

The compositional process which may be deduced from this source is more straightforward than that outlined above, to fit Matthew's sketch to Geoffrey of Monmouth. Rather than *Fossa*, Henry of Huntingdon begins with *Ikenild*, in precisely the orthography Matthew Paris likewise uses to name his straightest, shortest, and surest-drawn route. Above we noted that in graphing this vertical line Matthew follows the precedent of his single-itinerary maps; still more noteworthy is that each segment of the London-Apulia Itinerary begins not at the top of the page but at its bottom. The passage to Apulia and ultimately Jerusalem thus has, appropriately, an overall eastern orientation, just as *mappaemundi* do. But Henry of Huntingdon notes explicitly in his account of Britain's four great roads that Ikenild, the first route, passes rather in the opposite direction, from *orient* to *occident* (cf. the *Scema*'s parallel note that Ikenild leads *ab oriente in occidentem*). Matthew's sketch of Britain as multidimensional/multiple-route territory thus begins, precisely as a single-itinerary would, at the bottom-center of the half-page allotted for this figure. The other lines, whose courses, names, and toponymic orthography also follow Henry of Huntingdon very closely, fall into place relative to this first road, and hence again the curious feature of the *Scema*'s western orientation emerges as a direct result of Matthew's faithful graphic rendering, whether from a book before him or more likely from the book of his memory, of the precise disposition of material in his textual source or sources.[40]

In short, whether in beginning with *Ikenild Strete* at his half-page's bottom (per the text of Henry of Huntingdon and the spatial model of single-route cartography), or with *Fossa*'s source at *Totnes* in his page's upper-left (per the text of Geoffrey of Monmouth and the spatial models of scribal and illuminatorial *mise-en-page*), Matthew Paris displays, with an instructive immediacy in this "rude" and critically dismissed sketch, the precise trajectory of that conceptual leap so rare in medieval times—between the reception of geographical information in a textual format and its subsequent expression in figural or cartographic terms. Without question, every one of Matthew Paris's other maps is more polished, complete, or comprehensive than this. But none tells us as much about the St. Albans chronicler's striking textual-to-figural turn of mind, or about the scholarly practices and discourses from which any "cartographic revolution" in the thirteenth century must have taken its impetus.[41]

Whatever the process of the *Scema Britannie*'s composition, its feature of a western orientation makes one more thing very clear. Gone here, for the first time on any map of this territory, is the sense of a dominant pull eastward, whether to Rome or to Jerusalem. In place of this is a sense of territorial wholeness and of formal completion, an almost sacred quality and one which has been lent to the Figure of Britain on one hand by the chronicle tradition which informs the image both structurally and ideologically, and on the other by the formal character of other medieval map genres, whose centered, circular, and radial geometry this sketch invokes. Matthew Paris's maps have been said to be "remarkably free from the general medieval tendency toward symmetry"; but to its great effectiveness as an image, free from formal symmetry this *Scema* most certainly is not.[42]

The Gough Map of Britain has been seen for the most part in isolation, as a kind of spectacular, anachronistic deviation from not only the forms but also the functions, and the ideological concerns, of contemporary cartography in England. But such spatial and political issues as mark Matthew Paris's *Scema Britannie* also inform the Gough Map crucially. And indeed Gough—large-scale, compiled at considerable effort, and apparently a product of official or direct Westminster patronage in a way that Matthew's *Scema* is not—takes these themes a good deal further.

THE IDEOLOGY OF A ROAD MAP

Figure 3.1 reproduces the Gough Map of Britain, in a facsimile version more legible than the photograph of the original document appearing as figure P.2, above.

Compared with Matthew Paris's skeletal, 6 x 9 inch Figure of Britain, the Gough Map is a far larger (approximately 2 x 4 feet), much denser, and immeasurably more complex image, both geographically and iconographically. Where Paris's maps appear in multi-functional codices (primarily chronicles and sketchbooks), Gough for its part shows no signs of having been bound (within however large a folio), covering, rather, two large skins of vellum stitched together; by contrast the Hereford Map runs 5′2″ x 4′4″ on an extraordinarily large single skin. A series of nail-holes around its edges suggests that like Hereford (which occupied the central panel of a triptych at some point), Gough too may once have been fastened "on to a board or wall."[43]

These details, together with the internal evidence of much attention to contemporary travel systems, have led modern observers to suggest that Gough may have been designed as "an official map of Great Britain." Its service as a model for maps used (if not commissioned) by English government agents well into the sixteenth century—when it was superceded, finally and definitively, by another map associated with government patronage, Christopher Saxton's *Atlas* (1579)—would appear to bear this point out. And as it is further speculated that Gough "represents a type, of which several copies were in circulation," it has been taken as likely that such a document was kept at London (so E. J. S. Parsons) and/or at Westminster (so R. A. Pelham), deposited there "for all"—but especially royal couriers—"to consult."[44] This likelihood of a metropolitan display context, together with apparent official sponsorship of its compilation, makes the Gough Map particularly appropriate as the inheritor, although within an administrative rather than a Christological mode, of some of an earlier Westminster *mappamundi*'s associations and functions as an image of state and an emblem of kingship.

If the Peutinger Table has traditionally been considered "essentially" (Crone) or "solely" (Jervis) a road map, and the Matthew Paris Map of Britain "essentially" (Harvey) or "plainly intended to be" (Parsons) an itinerary map, it has been the habit of medieval map criticism to read the Gough Map in largely these same terms. Critics have gravitated toward such interpretations (of Gough as preeminently a practical aid to wayfaring) because this map's "main feature" has from the outset been taken to be its "remarkably detailed and accurate representation of the contemporary road system." The Gough Map has also long been praised for being "admirably accurate" in terms of its "overall depiction of the shape of the island." This depiction appears to owe its south and southeast coasts, its British Channel, and its North Sea to medieval portolan charts; these were waters with which Mediterranean ships were familiar. But Gough's accurate shape owes also to its roads. Many observers have been struck that the delineation of the island is best "where there is a bracing network of roads" and much the worse "where there are few or no roads shown"—as in Wales, Scotland, and southwest England.[45]

It is a curious testament to the durability of established interpretive channels that, just as he helped direct critical reception of the Hereford Map into the 1990s by suggesting its ancient use as an altarpiece, eighteenth-century antiquarian Richard Gough—for whom this map is named—likewise set out the terms according to which the Gough Map has been received. "The greatest merit of the map," Gough said, "is that it may justly boast itself the first among us wherein the roads and distances are laid down." Following up on this lead have been the Gough Map's most influential twentieth-century commentators, including R. A. Pelham, Frank Stenton, G. R. Crone, B. P. Hindle, and its Royal Geographical Society facsimile editor E. J. S. Parsons. No critic to my knowledge has yet dissented from Parsons's summary judgment in 1958: "From the detail shown, there is no doubt that its purpose was to serve as a map for travellers." Gough's links to "evidence bearing on medieval lines of travel" have been well documented.[46] Conspicuously unexplored, however, have been interpretations of the Gough Map along other, alternative lines, as well as consideration of such implications (e.g., social, political) as might extend outward from this consensus view of the road-system's central importance.

Matthew Paris, it will be recalled, centered his sketch of Britain's constituent Roman roads upon Dunstable, a town not far north of St. Albans; indeed the two sites are consecutive stops on the itinerary north to Newcastle that structures Paris's Maps of Britain. Paris has frequently been taken to task for this choice, by cartographic historians.[47] But although this sketch must be seen more as a paradigmatic than as a factual figure of Britain, Paris had good empirical basis for his placement of relatively obscure Dunstable at or near the center of his road system. For a number of Roman roads did in fact come together in the vicinity of Dunstable and St. Albans, among them the great highways of Watling Street, Ermine Street, and Akeman Street, with the Icknield Way not far removed and the Fosse Way connected if somewhat further off. Add to this the tendency of medieval documents to refer to any surviving local Roman road as Watling Strete or Fosse—"the names were often generic rather than individual," Ivan Margary has said, meaning "many of them were given to more than one road in quite different parts of the country"—and Matthew's sketch of the four famed highways' convergence not far from his own mid-island monastery seems amply justified. Like neighboring Dunstable, *Verulamium*, as the region around St. Albans was known in imperial times, was an area "intensively Romanized" and one which lay "in the heart of the Province"; its unusually thick-lying complex of roads may be a response to pre-Roman political geography, as the strongest tribe in Britain at the time of Caesar's invasion (54 BCE) had their center near St. Albans.[48] Such features of his local area as these, then (its many Roman roads, its rough physical and ancient political centrality), must have resonated for the cartographically minded historian, especially in light of the typological recommendation Paris had for the kind of centered and symmetrical structure as is his Figure of Britain.

In contrast to Paris's simple *Scema*, the Gough Map's far more elaborate series of roads "leaves no doubt" that in its compiler's mind London—not Dunstable—comprised "the centre from which the principal highways of England radiated." Indeed, as medieval road-system authority Frank Stenton has further judged, "whatever the limitations of his knowledge, the compiler of the map had at least arrived at the conception of a system of roads radiating from a national capital."[49]

But the Gough Map's centering of its road system upon London results in an overall composition which, if fuller and more detailed, is somewhat less balanced and pleasing, formally, than is Matthew's *Scema*. Gough's England is, like Matthew's figure, "approximately an elongated oval," as Parsons has observed.[50] But this body is not even roughly concentric or symmetrical, nature's ideal formal qualities (at least as maps and scholars such as Frederick II's set them out, respectively). Instead Gough is rather lopsided, weighted to the southeast. The flesh or matter of its geographical reality has grown deformed, as it were, in contrast to the *Scema Britannie*'s paradigmatic elegance.

Gough's southeasterly weight has, however, if no natural symmetry at least some justification in empirical and social terms. We noted in chapter one that London and Westminster took on greater and greater importance politically and commercially as the thirteenth century progressed, in gradual backlash to the decentered feudalism predominant in the eleventh and twelfth centuries (see "A World Map at Westminster"). More relevant still may be that while Roman roads radiate from a number of regional centers (note for example the spokes around Canterbury/*Durovernum*, tribal capital of Kent), nevertheless the imperial highway system in Britain had London (*Londinium*) as its main hub or "principal centre" (see fig. 2.11, top and right). The extent to which a Roman road system survived into the Middle Ages—as usable infrastructure, not just the overgrown marker of ancient imperial dominion—must remain partly a matter of conjecture. Yet in general these roads "had indeed been so well constructed that even in spite of neglect they were tolerably serviceable" into the thirteenth century. This point is borne out by the frequent correlation of Roman roads both with routes on the Gough Map and with those indicated by royal itineraries or constructed from commercial traffic information.[51]

Statutes interpolated during the reign of Henry III into a third version of the so-called *Leges Edwardi Confessoris* or "Laws of Edward the Confessor" provide for the basic maintenance of, and for extra protection from assault upon, four highways running from coast to coast and distinguished as *chimini regales*, "king's highways" or "royal roads." In early English law, as J. S. F. Tatlock has observed, there is "plenty about sanctuary . . . and about protection of the roads." Thus while the statutes in question were taken from the *Leges Anglorum*, composed c.1200–10 in London, they appear to have an earlier source or cognate in the *Leis Willelme* or "Laws of William" (i.e., the Conqueror), first compiled in the time of Geoffrey of Monmouth and Henry of Huntingdon, which hearken back in turn to the authority, ultimately, of pre-Conquest days. Since Edward the Confessor's name was already "one to conjure with," England's early

Norman Kings contented themselves with confirming, with additions, the laws of previous rulers. But while early versions of the names Watling Street, Ermine Street, Fosse, and Icknield Way appear individually in various Anglo-Saxon charters, their collection into the trope of Four Ways specially protected by the king's peace is an innovation which, though assigned to the Confessor, belongs apparently to the eleventh century and which by the twelfth had come to enjoy a measure of currency.[52]

For the law-codes' "four Royal public Roads," of course, are precisely those we have encountered already in the chronicles and in Matthew Paris's sketch. Their privileged distinction as royal highways, carriers of the *pax regis*, meant that offenses committed along these roads were to be tried before officers of the Crown, apparently to put them beyond the threat of local influence. All other, lesser roads (those "leading from one city to another, and from borough to borough, and along which merchandize is carried and other business done") were understood to lie rather under "the law of the county" around them, and were subject to the usual local customs and authority. A marked increase in trade and hence road-traffic, however, meant that by the early twelfth century the protection of the king's peace had been extended to include all the realm's highways or principal thoroughfares.[53]

This development looks to be extremely significant, in regard to the concept ascendant in the thirteenth century of an English regnal body as territory inscribed, and described, by a national network of roads—especially in that maps and roads are throughout both their histories inextricably linked. The thirteenth century's explicit attribution to St. Edward in particular, of antique laws giving special protection to travelers along a set of royal highways—spokes which sketch the coast-to-coast bounds of a British king*dom*, or that territory upon which a king's "doom" or judgment runs—is a cultural fiction which only reinforces the royal patron saint's basic corporeal association with Westminster, with an increasingly fixed center for the English monarchy, and ultimately with the reigning king's body. In much the same sense as those quasi-heraldic maps known to have been placed here, the royal highways fanning out from a metropolitan hub come to serve themselves, then, as emblems for the increasingly centralized monarchical state, the material inscription of its authority.[54]

Now, to be sure, such a map as Gough will have been useful to prospective medieval wayfarers (whether merchants, messengers of crown or church, courts on circuit, pilgrims, or others) and perhaps particularly to travelers on their way out from or back to London. But any map of a "[national] road system" will of necessity do more than simply help people get around—and this is the point at which commentary on Gough (as indeed on the Peutinger Table) has so far faltered, or at least chosen to stop short. Certainly, in one sense the Gough Map would appear to be a "map for travellers," and perhaps it is "primarily a roadmap of Britain" (holding aside, for the time being, its numerous other geographic and symbolic details). But what then is the ideology of a medieval road map? What is the social and historical meaning of this new kind of image of Britain?

The close and mutually constitutive relationship, historically, between roads and maps—particularly Roman roads and Roman maps—suggests that to determine the ideology of Gough as a road map, we ought to consider, in a basic sense, the ideology of Rome. Not just its maps but indeed the empire's roads themselves are ideological. As Margary has said, they constitute a "lasting monument to the Roman occupation" of Britain, and as we have seen they were not only a looming presence, but in use (especially by agents of central authority) well into the Middle Ages.[55]

A period of "gradual decline and breakdown" following the termination in CE 410 of imperial administration in Britain meant that by the thirteenth century Roman roads were largely piecemeal. But several other qualities of the road system in Britain stand out as definitive as well. First, Britain's Roman roads "form an impressively complete network." Although laid out by different surveyors and constructed at different times, still the roads comprising this network "give a striking indication of planning by a well-organized central authority." Roads "radiate from all the chief tribal capitals" but London, as we have seen, constituted "the main radial center" (see fig. 2.11, above).[56]

Concerning the physical characteristics of Roman roads, the principle of construction first observed is "a rigid adherence to very long straight alignments" (fig. 2.12), a quality which medieval chronicle-writers assign to King Belinus's roads as well. Second, the roads were provided with ditches along either side, and their gravel or paving-stones laid upon a high base (or *agger*) carefully constructed from this displaced earth and other local material, to ensure their proper drainage (fig. 2.13). Third and finally, because Roman roads were high-ways (i.e., on high ground) whereas springs and wells are on low ground, these roads' courses tend to be somewhat removed from village sites (as opposed to native British trackways, which typically run to and between them).[57] Each of these basic characteristics—their removal from the daily pathways of native life, their monumental size, and their striking straightness— is ideologically consistent with the functional historical observation that in their layout, Roman road systems link directly to a military purpose. They served, that is, first to aid in the conquest of new territory, and second to help police and administer it, after military domination had been effected.[58] The road network's notable completeness, its apparent central planning, and its radiation outward from a main London hub (accessible by water and on the near or empire side of the island, not coincidentally) all conform also to such a function.

In a point directly relevant for the Peutinger Table (see notes to discussion above) and perhaps indirectly for Gough, the Roman road system and map of same is believed to have had its origin in the transport system established by Augustus for moving officials and carrying mail, extended later to include troop movements. Closely linked was the *corpus agrimensorium*, a body of technical literature on the measurement of conquered land and its division into "hundred square units for distribution among Roman colonists"—a practice known as centuriation.[59]

Samuel Egerton has described the apparent need of the Romans to organize their world on the model or pattern of the grid, a "hierarchical design" of innate visual and administrative appeal and one based on the repetition of a single unit, or module, as the structural basis for the entire larger composition. "How monotonous," Egerton reflects, "must have been the centuriated landscape and gridded towns of Europe under Roman occupation!" Both geometrically and politically speaking, what is key to this concept is that it is centrifugal: "Like the imperial ambition it so well symbolized, the grid rolled out relentlessly in all directions from the capital, homogenizing everything in its path." In addition to simply moving mail and officials, then—and all the more so because Roman roads were so conspicuously monumental, and so monumentally useless to local communities—the territorial composition which a radial road system or its accompanying map-image may be said to comprise will in essence have served "as a surrogate army of occupation, constantly reminding subject peoples of their conquerors' superior organizing power."[60]

Yet the spatial dynamics of the empire's road system do not draw the traveler only outward. They also pull inward: all roads, even in the provinces, run eventually to Rome (both literally and by aphorism), meaning the system's magnetic charge is alternately positive and negative, its force both centripetal and centrifugal. The imperial traveler is called to the center even as pushed to new borders, toward the survey and domination of ever wider circles of territory.

Medieval maps which treat land travel are almost invariably based in part on the routes of Roman roads, if indeed they are not based directly or indirectly (as may be the case in Gough) on the model of a Roman road map itself.[61] The ideology of a medieval road map is hence inescapably an imperial one. Such a text's concern, like such a technology's concern, is first for the military acquisition of territory, and next for its administrative regulation and economic exploitation.

A MONARCHY OF THE WHOLE ISLAND

The loss of Normandy in 1204 meant that English kings thereafter tended increasingly (with some notable exceptions) to give their primary political and administrative attention to the British Isles rather than to territories overseas. It is a direct corollary to this tendency that "as a Westminster-based English governmental machine became increasingly complex and bureaucratic"—so R. R. Davies has said—"the possibility of extending its reach beyond the boundaries of England and of introducing its norms in legal habit, administrative structure, and financial accountability into other parts of the British Isles on a systematic basis began to be entertained—indeed, acted upon."[62] One governmental or clerical discourse intimately involved with such a multifarious imperial action was the cartographic—whose developments in terms of an English regnal image we have charted to approximately 1255–59. The next generation's prototype for the Gough Map, as we shall see in the following chapter, is far more than simply a map for travelers. Rather, such a map underscores Davies's linkage (through legal, administrative, and fiscal practices) of an increasingly centralized bureaucracy to an English imperial project within Britain—and beyond it.

The importance to England in this period of "internationalism" as a cultural and political factor (the terminology is Elizabeth Salter's) has often been underemphasized. Nonetheless, England's definitive "separateness" from the continent (in Michael Clanchy's view) had long been more than geographically apparent, and this quality—a "sense of Englishness" which was latent in the twelfth century but which emerges in the thirteenth as "a political force"—was underlined forcefully by the events of 1258–65. Clanchy has asserted that "this period of rebellion and civil war marked the turning point in the definition of English identity." In part this was because the Treaty of Paris (1259) ceded Henry's Norman and Angevin lands to Louis IX of France, in return as it were for Louis's support of monarchy against the Barons' revolutionary "Commune of England." By no means did Edward I and his successors disavow campaigning abroad, of course. But when England's royal house pursued foreign ambitions after the Barons' Wars, they did so now "as heads of an English nation" and with fiscal ratification by the community of the realm, such that the enterprise became a collective state project rather than a narrowly feudal one.[63]

The medieval conquest of Wales was undertaken not according to the terms, at first, of some overarching imperial plan or "dream of centralism." English (more properly Anglo-Norman) domination of the Welsh developed initially, rather, as a result of the disconcerted actions of an "acquisitive aristocratic lordship." It was begun, that is, by Norman warlords given land in the Welsh Marches after the Conquest, who were motivated individually to improvement of their material and political standing in a frontier, feudal economy, and who were able to carve out powerful fiefdoms while royal eyes were diverted by richer prizes like Normandy and Aquitaine. In the eleventh and twelfth centuries, then—what Welsh historian Davies terms the "pre-bureaucratic age" of Anglo-Norman/English imperialism in Britain and Ireland—the primary agents and institutions of domination were, in a word, baronial.[64]

This changed completely in the thirteenth century. Increasingly during Henry's reign and very decidedly during Edward's (under whose rule England was united, Powicke has said, "as it had never been before"), the tenor of the project instead became "royalist, England-centered, and uniformist." "Conquest" in the formulation of the mid-thirteenth century "now implies an increasing degree of administrative integration and central direction." From the monarchy's symbolic and administrative seat at Westminster, with its pointed display first of a heraldic *mappamundi* and later (apparently) of an official map of Britain, a "metropolitan view of authority," as Davies has put it, "was extending outwards to embrace all the king's land in the British Isles." On the horizon by 1300 was "the prospect of a single, unified, England-centered kingdom."[65]

This "metropolitan" view of space and authority in Britain—also fitting might be to call it an imperial and fiscal conception of English kingship—has a curious genealogy. We have seen that for Kantorowicz, sacramental kingship (such as Henry III's) had its essence in coronation, in particular the ceremony of consecration through which the anointed ruler took on a special resemblance to Christ and received a share in divine authority. The thirteenth century saw a gradual but marked devaluation in coronation's role as kingship's operative legal basis.[66] However, it is an intriguing footnote to this change that an apparent irregularity in the 1216 coronation of Henry III seems to have played a key role in the development of later kingship's new basis in territoriality and in the concept of an inalienable royal "fisc" whose bounds were coincident both with the king's communal symbolic body and with the rights of the community of the realm.

The last years of John's reign (1199–1216) were dominated by conflict between the king and his barons, who (as so often) were angered to the point of rebellion by the crown's "costly and abortive" foreign expeditions. In the course of their conflict both sides were driven to seek assistance abroad. Although England had been under Innocent III's interdict since 1208 (for its support of Guelph Emperor Otto IV) and although John himself had twice been excommunicated, in 1213 the king outflanked the barons by making his kingdom a vassal state to the papacy. For the next several years Papal Legate Guala Bicchieri, an Italian Cardinal, was a prime force in English politics. A rise in baronial fortunes led to John's forced ratification of Magna Carta in 1215 (although after negotiations it was reissued as a freely granted charter). When Innocent declared it invalid and excommunicated thirty barons, however, civil war broke out immediately—the baronage had kept itself in arms by calling frequent tournaments. At the invitation of the English barons, Prince Louis of France (afterwards Louis VIII, 1223–26) invaded in May 1216 and early on met with good prospects for success, despite his inevitable quick excommunication by Guala Bicchieri. However, support for Louis waned with the king's death in October 1216 (once the unpopular John was "no longer present as a personal focus for discontent"), and a year later Louis had been bought out of his claim to the throne.[67]

At the height of this conflict came the business of a nine-year-old Henry III's accession to a disputed throne. Papal Legate Guala occupied the power-vacuum and helped smooth the realm's transition into the long era of Henry's minority (1216–27). In a few years, the recalled legate would apparently bring back to Italy an English-designed *mappamundi* (the Vercelli Map, c.1215–18) in addition, it has been argued, to one of the four great manuscripts of Anglo-Saxon prose and poetry (the Vercelli Book, c.950–999). Guala Bicchieri transformed the struggle to set Henry firmly on the English throne into, Clanchy has said, a holy war. Royalist troops were accorded the status and took on the white cross of Crusaders, a point which seems to resonate with the legate's interest in *mappaemundi* and apostolic literature and which may also inflect the slowly increasing status of native English literary efforts. Because Louis and the barons held London, Henry's coronation took place at Gloucester, and because the Archbishop of Canterbury and the crown and other regalia were not available, it was a relatively simple affair, with the papal legate presiding.[68]

By all accounts Henry swore the tripartite oath customary from Anglo-Saxon times. According to the dovetailing arguments of Kantorowicz and H. G. Richardson, however, Henry's 1216 coronation oath also included a new, fourth clause—"in which the king promised not to alienate rights and possessions of the Crown and to recover what had been lost." Such a promise had been unknown as late as John's coronation in 1199, but though it remained uncodified Edward I would refer to it no fewer than eight times during his reign, to justify various grabs for territory.[69] The role of Bicchieri in this addition of an explicitly *territorial* element to the sworn duties of English kingship is crucial because in administering the young monarch his oath of office (so Kantorowicz argues), it would seem that the legate merely followed, for the action of royal investiture, a practice familiar to him from ecclesiastical investment ceremonies. An extra or additional "eighth clause" concerning inalienability, not recorded in Canon Law but known to have been employed because of references to it by glossators, was frequently added in the twelfth and thirteenth centuries to the standard seven-part oath in the ceremony of episcopal consecration at Rome. Following canonical procedure, then, Bicchieri duly added to the usual tripartite English royal oath an extra, comparable non-alienation clause whose absence from coronation texts "[need not] startle us, for the corresponding clause was absent also from the standard [episcopal] oath."[70] In short, just as certain new bishops swore (in an extra non-codified oath) not to alienate property entrusted to them with their office, Bicchieri went beyond the codified form of the traditional English coronation oath in requiring Henry to forswear alienation of any of the property or rights entrusted him with his office.

The uncompromised maintenance of the lands comprising the English kingdom took on greatly increased importance for Rome following 1213, with John's symbolic surrender of the realm to the papacy. Richardson takes this as evidence suggesting that Henry III's "additional promise" may have been introduced quite consciously by the papal legate, where Kantorowicz's version of events (above) imputes the appearance of the new clause to a combination of procedural confusion and half-conscious conceptual migration.[71] Either way, this transfer to the secular realm ("the Crown") of the *ecclesiastical* concept of non-alienation marks one further step by which the monarchical state moved in the thirteenth century toward a bodification comparable to the church's—indeed, toward a corporational identity set directly against it.

We have noted that in the early thirteenth century a series of interpolations were made into the *Leges Edwardi Confessoris* of material from the *Leges Anglorum.* This resulting third version of the Laws of Edward the Confessor bears testament, Richardson has noted, to the popular currency of the notion that a king was obliged to maintain the Crown's rights and territories unimpaired and indeed "should swear to do so at his coronation," in just such an additional promise as Henry appears to have made at Guala's initiative. St. Edward's Laws were fairly widely known in this new, interpolated version, to lawyers, king's clerks, and some historians (although by the latter they were more often encountered indirectly). That Henry Bracton knew them is "beyond doubt"—indeed his own work transmits "the theory of kingship embodied in the Laws . . . substantially unaltered."[72] Thus the concept of an inalienable royal fisc (which Bracton explores in detail) helps the emergent English state to take on, following the ecclesiastical model, an identity at once territorial, social, and corporate. The extension of this concept not long after the mid-century, as Davies has observed, "to include the Crown's lordship and estates in Ireland and Wales," helps bring to full articulation and momentum an ideology of English conquest—or more precisely, re-conquest—in Britain.[73]

The correct term here is "re-conquest" because Henry and Edward swore not only to maintain all the rights and possessions they inherited but also to recover those which had been lost. In the imagination of thirteenth-century English kings, clerks, and chroniclers, England's ancestral lands properly included all of the (so-called) British Isles, and more—Norway for example and even Rome itself: in short all those territories which legendary kings such as Brutus, Belinus, and especially Arthur had held and which their successors were to hold again.

There is further good reason, beyond Guala Bicchieri's presence at Gloucester in 1216, that a non-alienation clause should first appear in the English coronation oath in the early part of the thirteenth century; for these were also years which saw the *Leges Anglorum* composed in London. This anonymous treatise, as we have seen, purports to treat and even "reflects some knowledge of law." However, its text is infused more importantly with the "glamourous ideals" of Geoffrey of Monmouth, whose conjuration of Arthurian legend included as its central territorial proposition and political focus the concept of a *totius insulae monarchia* or "monarchy of the whole island," the King of England's ancient hereditary right to overlordship in Britain.[74] Most of the individual quests of medieval English Arthurian literature may be read to some extent as strands in this overarching theme, of Arthur's glorious and inexorable extension of an ideology of unity and centralization in Britain around the verge of a stay-at-home patron-king, his court Camelot, and its recombinant Round Table of far-flung knights and petty British princelings.[75]

But the *Leges Anglorum*'s direct importance to Henry's 1216 coronation lies, more than in its Arthurian ideology, in its compiler's allegation that good King Edward had at his accession sworn precisely such an oath as Henry's, first to recover England's previously alienated lands and possessions, and thereafter "to observe and defend all the dignities, rights, and liberties of the Crown of this realm in their wholeness." During what has been characterized as the largely pre-"articulate" era of Anglo-Norman legislation before Magna Carta, the Laws of Edward the Confessor (however "vague") served as the realm's only compilation of legal principles or "criterion of good government," afterwards lingering as a kind of popular legendary counterpart to the often reissued great charter.[76] What is most striking for our purposes here, however, is that the *Leges Anglorum* tie the concept of a "monarchy of the whole island" taken from Geoffrey of Monmouth's Arthurian conception of British history together with the thirteenth century's authorizing sacral and royal figure of St. Edward, whose grant of higher jurisdiction to a set of four royal roads (at least in popular attribution) serves to sketch upon what is otherwise a feudal landscape the skeleton of a centralized state bureaucracy.

Another text which links these two legendary, prototypical kings (Arthur and St. Edward) directly to late thirteenth-century royal ideas about English territory and history is the *Flores Historiarum*, the St. Albans universal chronicle into which Roger of Wendover (c.1204–1234) incorporated much material from insular histories (including Geoffrey of Monmouth's and Henry of Huntingdon's) and which Matthew Paris adapted and extended (for the years 1234–1259) into the *Chronica Majora*. In the absence of a historiographical tradition of their own, the house of Benedictines at Westminster adopted this work from St. Albans and extended it themselves from 1265 through the end of the reign of Edward II in 1327. Not surprisingly, considering that their rise in fortunes during the thirteenth century coincided with Henry's promotion of the Shrine of St. Edward as the material and symbolic center of the English monarchical state, the Westminster Benedictines continued the *Flores* with what has been called a "pronounced royal bias"—this as opposed, for example, to Matthew Paris's frequently critical stance toward Henry III and Roger of Wendover's harsh view of John.[77]

Westminster manuscripts of the *Flores Historiarum* were decorated lavishly, principally with coronation scenes. These images mark the traditional insular chronicle division of English history into sections characterized by the reigns of notable native kings—in secular reprise of the standard ecclesiastical division of history into the eras of Adam, Noah, David, and so forth (as seen in illuminated Bibles and Psalters and in the work of universal chroniclers including Peter Comestor and Honorius of Autun). The most lavish sequence of coronations—there are ten, beginning with King Arthur's (fig. 2.14) and ending with Edward I's (fig. 2.15)—appears in the Chetham Library manuscript of the *Flores*, begun by Matthew Paris himself but illuminated at Westminster beginning in the 1250s. The Chetham image of St. Edward ("inevitably the richest" of the *Flores* set) has been linked to the great mural added in the mid-1260s as the Painted Chamber's centerpiece, since this monumental work presents "a very similar version" of the Confessor's coronation (compare fig. 2.16 to fig. 1.2, above).[78]

The fact to be borne most in mind with the Chetham coronation scenes is their generic nature overall: they are "in no sense documentary" but are rather, as Binski observes, essentially manifestations of a "standard image type."[79] This means that legendary British overlord King Arthur, lawgiver and sacral interlocutor St. Edward, and their contemporary inheritor King Edward I come with all the rest of the ten to blend, in essence, into a single royal image extended across the text of British history (but pushing outward, via an imperial destiny, to a degree of universal prominence). Crowned in each scene is thus a recurrent version of one dynastically authorized body (never mind the details of a broken line of descent), an English monarch at once individual and collective whose overlordship of Britain (and leadership of Christendom) is a matter of sacred, hereditary, and inalienable right.

Edward I's assertions of authority in Wales came intermittently from before his accession (he inherited all the Crown's estates in Wales and Ireland in 1254) through into the 1290s, but his major campaigns took place in 1277 and 1282/83. With the latter of these Welsh wars the principality lost what political and administrative independence it had gained in the early-century advent of Llywelyn the Great, and future armed conflict (e.g., 1287, 1294/95) had rather the nature of rebellion and its suppression.[80] The end of Welsh independence was marked by the execution of the last native-born Prince of Wales, Daffyd ap

Gruffudd, in 1283. A shattering of sovereignty, this event was inscribed territorially (much as the death of Llywelyn ap Gruffudd had been in 1282 and those of Scottish insurgents William Wallace and Simon Fraser would be in 1305 and 1306) by the placement of the Prince of Wales's head upon a stake at the Tower of London and the removal after drawing and quartering of the four parts of his body to four corners of the realm. In this act the aspiring conqueror seems to play, in grotesque inverse, upon the concept of the body of a prince as coincident with the bounds of his realm—or rather, as manifestly no longer such, in the case of conquered Wales.[81]

There is evidence that in the mid-1280s Edward I made plans to mark the achievement of his conquest of Wales with a reenactment of his coronation ceremony. The practice was common enough: both John (1199/1203) and Henry (1216/1220) had second coronations. The difference in this case was that Edward intended to play a role, that of King Arthur, apparently in extension of the tournament or Round Table celebration he held at Nefyn in 1284. What may have put the king in mind of this was his recent acquisition of the Crown of Arthur, one of Wales's greatest relics and a sign of its sovereignty (eventually it was removed to Westminster and attached to the Shrine of St. Edward, just as Scotland's coronation Stone of Scone would be in 1296). Edward's plan was to use this crown in his ceremony. By taking his legendary predecessor's place in a coronation tableau, just as he does (effectively) in the Chetham *Flores* series of coronation images, Edward asserts explicitly a direct bodily identification with King Arthur, in a stance not unlike Henry III's with regard to St. Edward.[82]

An association between English kings and Arthurian legend may be traced back to Geoffrey of Monmouth's dedication of his *Historia Regum Britanniae* to King Stephen in 1136. Yet as R. S. Loomis has observed, no English king before Edward I can be described as having had "more than a passing interest" in Arthuriana. Loomis attributes the greater part of Edward's interest to "sentiment" and "simply to the fact that he was a man of his time." But others have emphasized (and the above should underline) that political calculation informs much of Edward's Arthurian pageantry. Elizabeth Salter has noted, in particular, the frequent alignment of "royalist causes [with] Arthurian themes." This alignment, and its flip side—that baronial patronage is in this period in England almost never associated with Arthurian literature, or (so far as we know) with cartography for that matter—has important repercussions for the coincident development we have been exploring, in royal conceptions of kingship and in regnal mapping practices throughout the thirteenth century.[83]

Two further events will help characterize Edward's Arthurian enthusiasm—both of them connected with his campaigns in Wales. The first occurred during a visit to Glastonbury in the spring following Llywelyn ap Gruffudd's submission to terms, in November 1277: at Easter, Edward ordered the tomb of Arthur opened, probably to display his dynastic connection with this most authorizing figure in the history of British kingship, but also, apparently, to prevent the Welsh from rallying behind the prospect of their native hero's promised return. The second event came in 1301, when King Edward presented to his son and heir (the future Edward II) a copy of Geoffrey of Monmouth's *Historia*. Edward of Caernarvon had been born during the royal household's extended residence in Wales in 1284, just a few months before the king marked his conquest with various Arthurian festivities. Significantly, young Edward received the *Historia* upon the occasion of his confirmation as first English Prince of Wales; and given his father's political use of this text in precisely this year (on which more later), perhaps Geoffrey's mythography was intended as a kind of handbook to his line and rights.[84]

The concept of a monarchy of the whole island centrally informs Arthurian territorial ideology, as we have noted already. But Arthurian mythology has further implications for Edward I's conception of the spatial duties of English kingship. This literature, and in turn the literature upon it, is extremely vast; quality recent work has thus stressed, accordingly, that there is "not a single, monolithic Arthurian legend, but rather a multiplicity of Arthurian legends," even very early on. The various geographies operative for medieval Arthurian legend, however, are not just multiple and shifting text to text, across various genres, languages, and historical contexts; instead, they are frequently in conflict within even a single work, competing with one another for ideological primacy.[85] Still and all, certain generalizations are possible.

In a study of the "symbolic geography" of Arthurian romance, Rosalie Vermette has observed that a simple inside/outside or center/periphery model best characterizes the medieval Arthurian conception of space. Just as is the case in England in the thirteenth century, "the center of the Arthurian cosmos is the king's court, with the absolute central point being the king himself." The conventions demand, typically, a royal agent's departure from "the center of the court circle, the safe and secure known world," on a radial journey outward. The further a Knight of the Round Table can penetrate into the white spaces beyond the king's law, across the concentric circles of court, kingdom, and known world's edges toward the territory outside (the site of "chaos, disharmony, and injustice"), the more successfully executed has been the western Christian project of establishing a centralized political, social, and religious order.[86]

Vermette identifies Arthur's Round Table as this world's central image. And insofar as "the geometric figure of the circle is a symbol of wholeness and harmony," the Round Table reflects "the state of order and harmony that characterizes the Arthurian universe." Vermette also notes the direct connection, in Arthurian ideology, between a king's body and his kingdom's lands: the legendary Fisher King's "bodily handicap and sterility," for example, is "mirrored in the economic, social, and moral sterility of his kingdom."[87]

During the thirteenth century Winchester Castle's Great Hall (built c.1222–1235) housed both a painted *mappamundi* commissioned by Henry III in 1239 (now lost) and a great wooden Round Table, many times repainted but apparently connected in the first instance with Edward I's interest in King Arthur. Although fitted at one point with legs, it now hangs (and may originally or intermittently have done) on the wall (fig. 2.17). Although the Winchester Table recently has been subjected to thorough examination in all its aspects (especially archeological), it remains impossible to say what exactly the initial composition may have entailed. However, in its final, c.1521 repainting (fig. 2.18), the eighteen-foot table depicts an orb-bearing "Kyng Arthur" enthroned atop a concentric Tudor rose which forms the panel's hub—with the twist that "Arthur" here closely resembles King Henry VIII, down to the detail of his recently grown beard. In a wheel around the table from the king, beginning with "Sir Mordrede" at Arthur's right hand and "Sir Galahallt" and "Sir Launcelot deu Lake" to his left, the sieges or assigned places of 24 knights are inscribed, and extend radially from edge to center in alternating sectors of green and white. The Table's central emblem, white rose upon red, signifies the Tudor unification in a new badge and political order of the House of York's white rose with the House of Lancaster's red one.[88]

The late date of this repainting, of course, makes for a certain difficulty in interpretation. Still, in its depiction of King Arthur/King Henry at its center and head, together with its twenty-four spokes marking knight-sieges which are at once centripetal and centrifugal, the Winchester Round Table makes it apparent that the basic components of early Arthurian geography have their counterpart in the geometric and ideological qualities of thirteenth century territorial maps. That is, just as the Arthurian cosmos is ordered by its centered (Arthur/Camelot), circular (Round Table), and radial (quest-oriented) social model, and further equates the physiological body of the king with the territorial body of his kingdom, so, similarly, may be characterized the spatiality of medieval *mappaemundi*, *situ Hierusalem*, and (from the *Scema Britannie* onward) maps of Britain. Like Winchester's Round Table, their space is centered, it is circular, and it is radial. And finally, metropolitan space, the space of all these medieval map-types and of Arthurian literature, is defined by a corporate association, one both social and territorial and one whose basic identity is lent it by the figure of its symbolic and political head. Just as Christ is living Host, Holy City, and Incarnate World, and is the basic stuff of (not just an image on) the thirteenth-century *mappamundi*, so too, in Arthurian mythology and increasingly in English royal ideology and regnal cartography, does the king become much the same thing as his kingdom, the matter of the one ("matter" in all senses) indistinguishable from the matter of the other. Doubling Edward I's body, in short, is the territory covered by his authority.

It is a caveat among the most famous in the discipline of geography that "the map is not the territory."[89] For late thirteenth-century England, however, something almost directly opposite might be asserted. Per the fond ambitions of royal ideology at least, the map is the king is the territory. And if, as Kantorowicz has argued, in the budding political theology of the age the English king has two bodies, one "feudal" and the other "fiscal," the latter of these—as chapter three will set out in detail—is that which, effectively, the Gough Map of Britain describes.

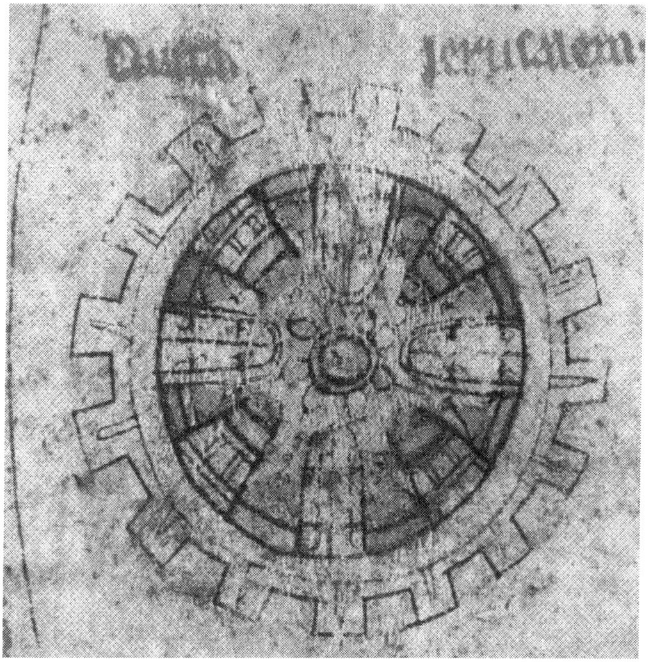

Figure 2.1. The City of Jerusalem, detail from the Hereford Map; reproduced by permission of the Dean and Chapter of Hereford and the Hereford Mappa Mundi Trust.

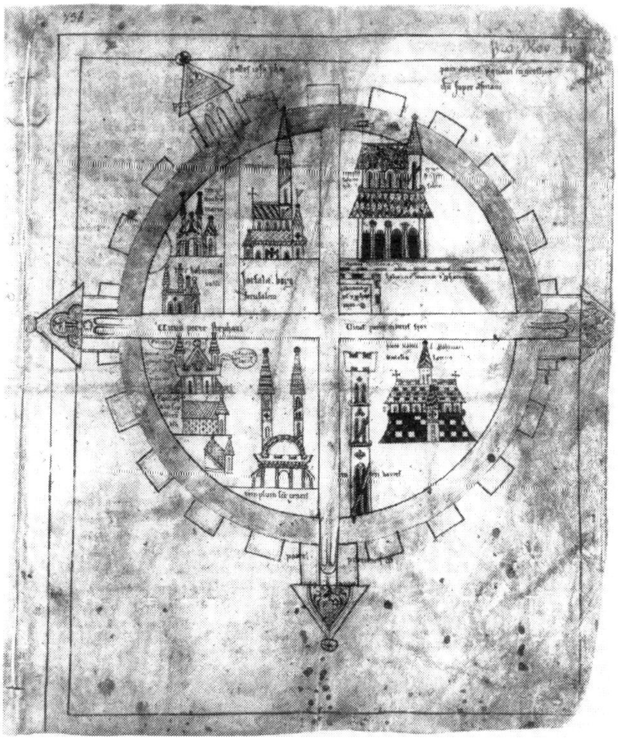

Figure 2.2. Map of Jerusalem; Copenhagen, Det Arnamagnaeanske Institut, AM 736, I, 4to, fol. 2; fourteenth-century copy of c.1180 prototype; note battlements and heavy cruciform emphasis, with resulting four-part division of the city; reproduced by permission of Det Arnamagnaeanske Institut.

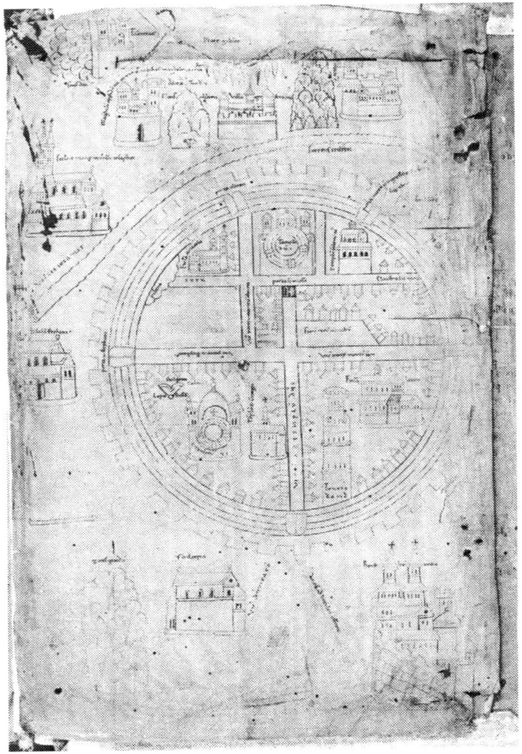

Figure 2.3. Map of Jerusalem, Stuttgart; Württembergische Landesbibliothek, Cod. Bibl. Fol. 56, fol. 135; note battlements and overall clarity of cruciform scheme (despite additions in top half), plus prominence of Templum Domini in circular shall above T-bridge; reproduced by permission of Württembergische Landesbibliothek and Fotoarchiv Marburg.

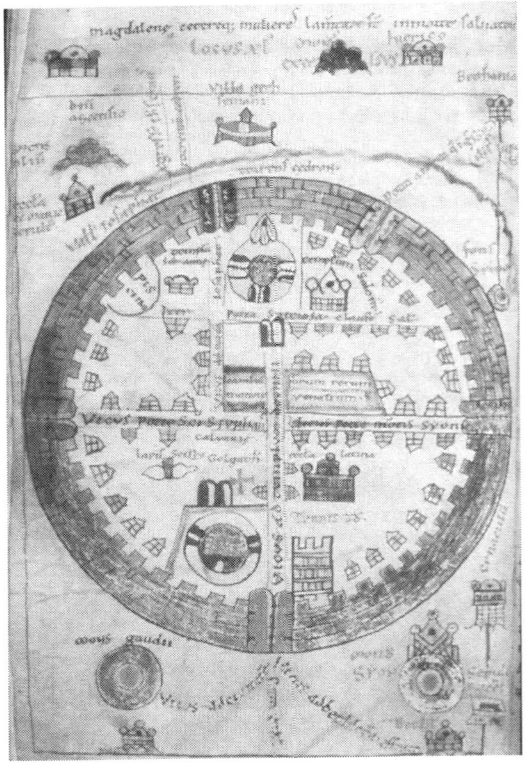

Figure 2.4. Map of Jerusalem, Saint-Omer; Bibliothèque Municipale MS 776, fol. 50v; note block walls and battlements, cruciform scheme, and circular Templum Domini surmounting schematic T drawn by avenues/thoroughfares; reproduced by permission of Bibliothèque Municipale, Saint-Omer.

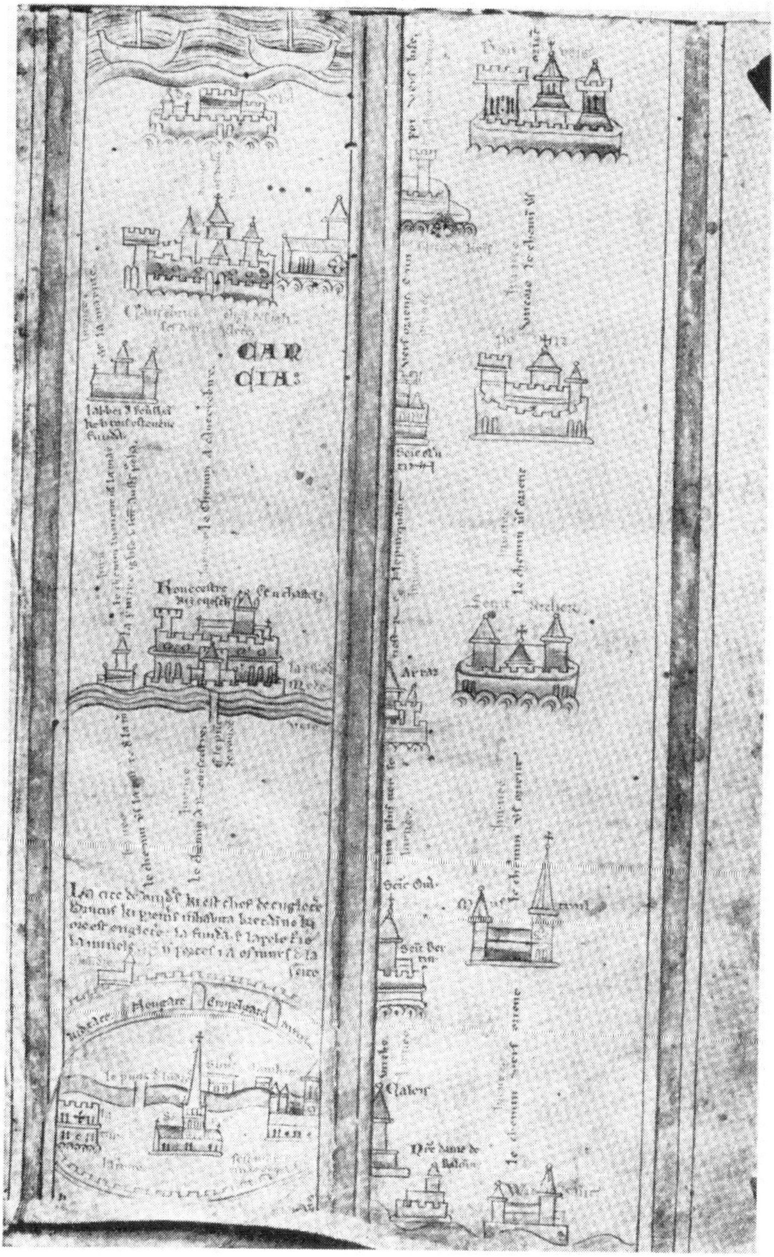

Figure 2.5. London-Beauvais, from Matthew Paris's London-Apulia Itinerary; Cambridge, Corpus Christi College MS 26, fol. i; St. Albans, c.1250–55; note the elaborate oval view/plan of London; reproduced by permission of the Master and Fellows of Corpus Christi College, Cambridge.

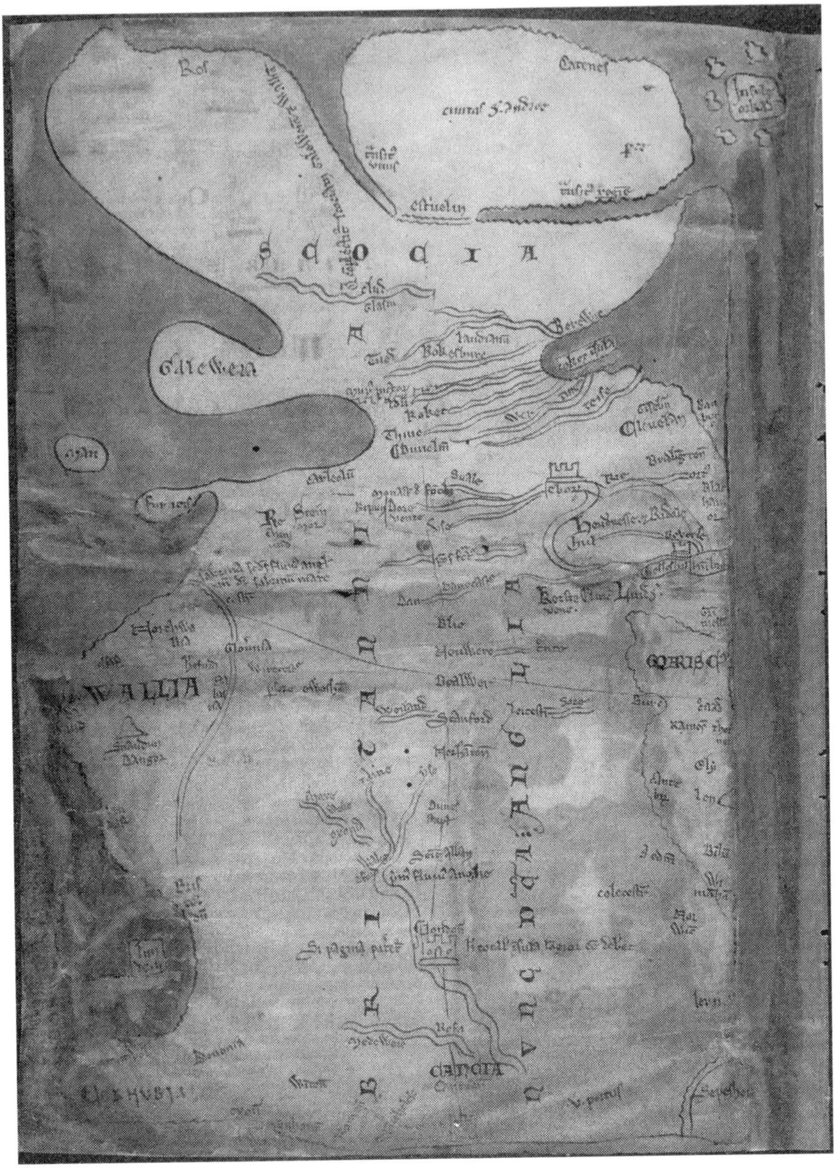

Figure 2.6. Map of Britain, Matthew Paris; London, B.L. Royal 14.c.vii, fol. 5v; St. Albans, c.1250–55; note structural prominence of Newcastle-Dover route; "either an early effort soon discarded or a late, very rough revision never completed"; reproduced by permission of the British Library.

Figure 2.7. Map of Britain, Matthew Paris; London, B.L. Cotton Claudius D.vi, fol. 12v; St. Albans, c.1250–55; note density of text and much increased attention to topographic details, especially coastlines and rivers; Newcastle-Dover itinerary (complete with town-icons) still very prominent as backbone; reproduced by permission of the British Library.

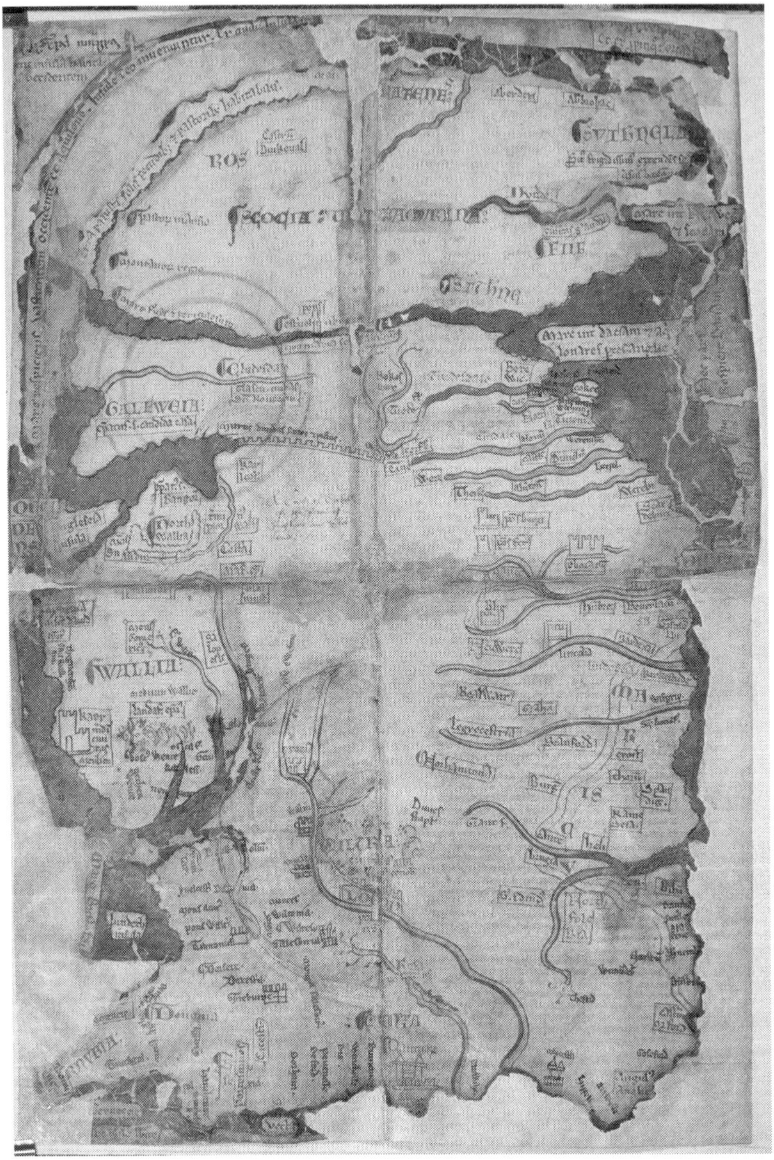

Figure 2.8. Map of Britain, Matthew Paris; London, B.L. MS Cotton Julius D.vii, fol. 49v; St. Albans, c.1250–55; given similarities to the finished (if fragmentary) Corpus Christi version, the rougher Julius "may represent a first draft that was later discarded"; the map was bound originally within the historical miscellany of St. Albans infirmarer John of Wallingford (note the creases), who added many of the place-names in his own hand; reproduced by permission of the British Library.

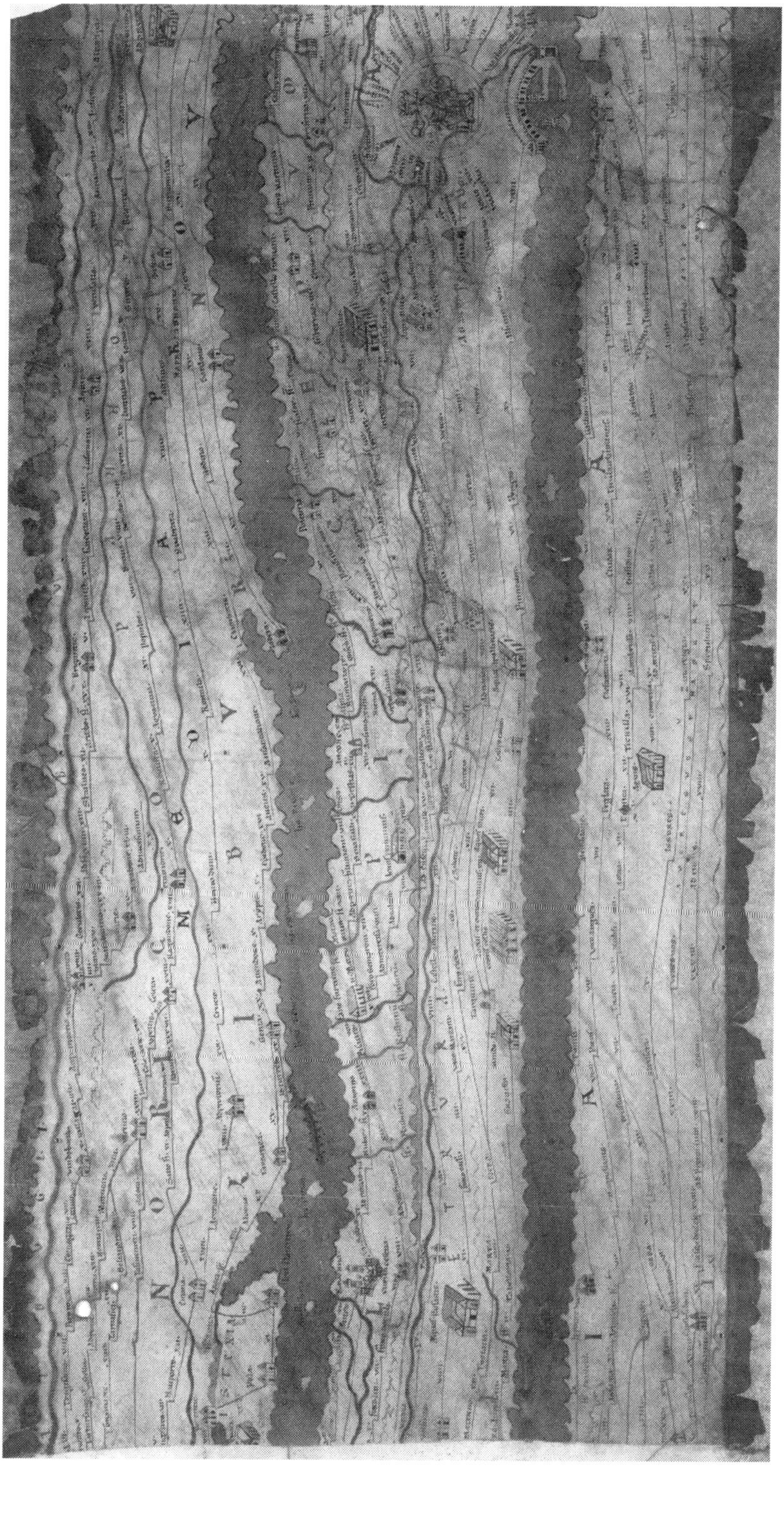

Figure 2.9. The *Tabula Peutingaria* (Peutinger Map or Table); Vienna, Österreichische Nationalbibliothek, Codex Vindobonensis 324, Segment IV; copied c.1230, late-antique prototype; note lines for routes extending outward from Rome; reproduced by permission of the Bildarchiv d. ÖNB, Wien.

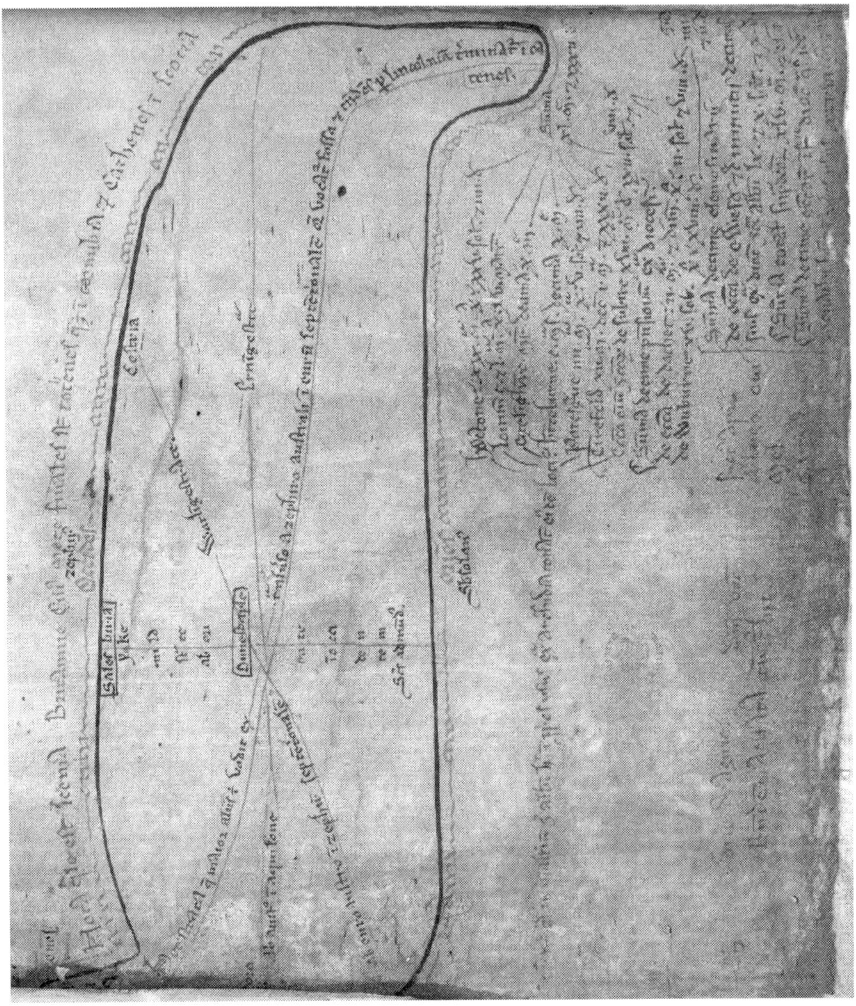

Figure 2.10. The *Scema Britannie* or "Figure of Britain," from Matthew Paris's *Liber Additamentorum*; London, B.L. Cotton Nero D.I, fol. 187v; St. Albans, c.1255–59; reproduced by permission of the British Library.

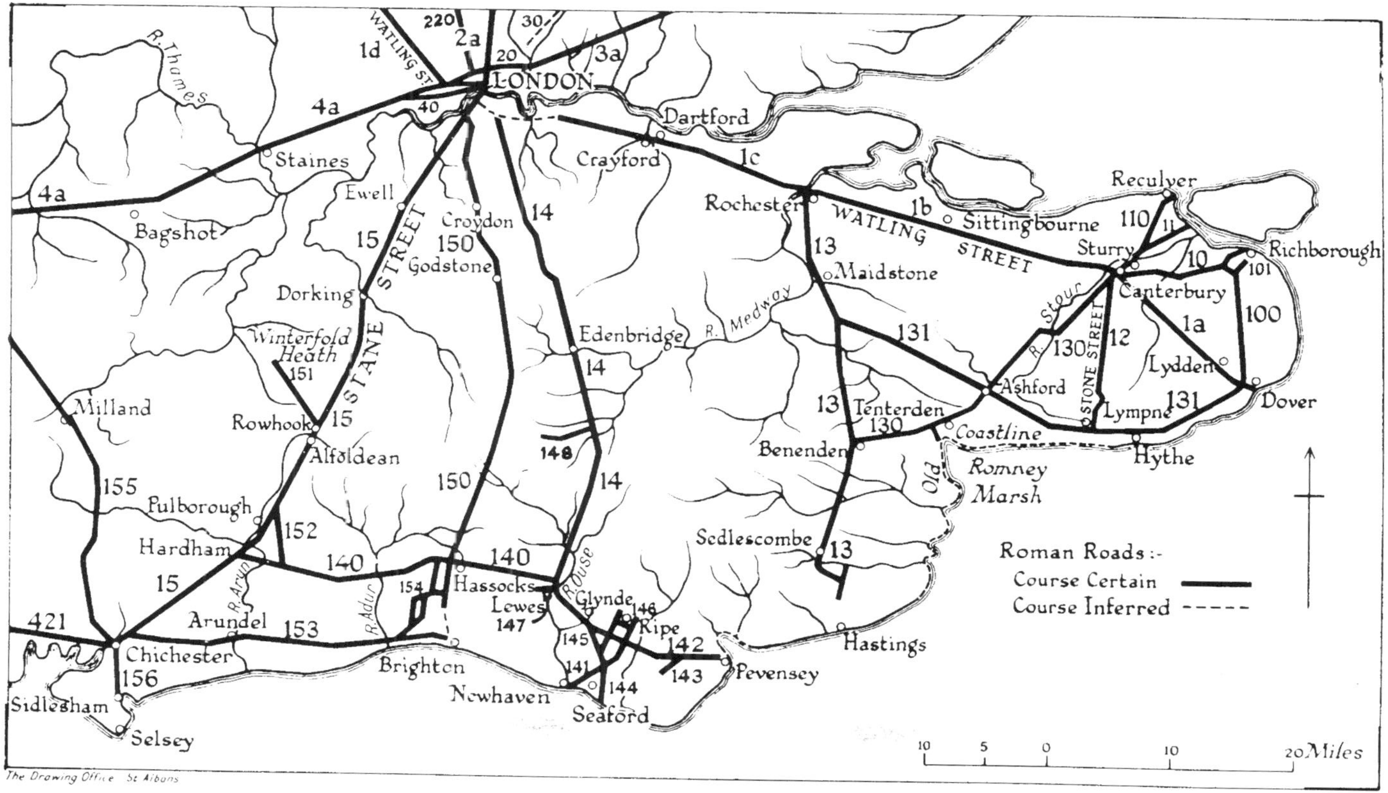

Figure 2.11. Britain's Roman Roads: London as principal hub, Canterbury as regional hub; diagram from Ivan Margary's *Roman Roads in Britain* (London, 1955), map 1; Phoenix House.

Figure 2.12. Fosse Way (near Easton Grey, Glos.), route marked by field enclosures; from Ivan Margary's *Roman Roads in Britain* (London, 1955), frontispiece; Phoenix House.

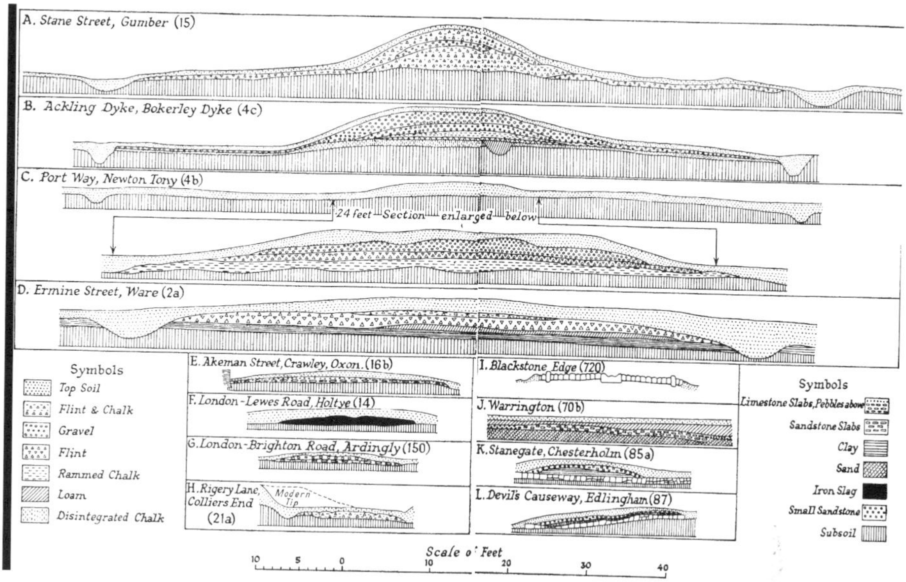

Figure 2.13. Diagram showing typical road sections; from Ivan Margary's *Roman Roads in Britain* (London, 1955), pp. 500–501; Phoenix House.

Figure 2.14. Coronation of King Arthur, Chetham *Flores Historiarum*; London, B.L. MS Loan 94 (formerly Manchester, Chetham's Library MS 6712), fol. 53; c.1250s-65; reproduced by permission of Chetham's Library.

Figure 2.15. Coronation of King Edward I, Merton *Flores Historiarum*, Windsor, Eton College MS 123, fol. 237; late thirteenth-century (i.e. produced after figs. 2.14 and 2.16); this image is a fuller and more accomplished stand-in for the closely related but less legible version appearing in the Chetham *Flores*, London, B.L. MS Loan 94, fol. 247v reproduced by permission of the Provost and Fellows of Eton College.

Figure 2.16. Coronation of St. Edward the Confessor, Chetham *Flores Historiarum*; London, B.L. MS Loan 94 (formerly Manchester, Chetham's Library MS 6712), fol. 115v; c.1250s-65; compare to fig. 1.2 (Painted Chamber Coronation of St. Edward), above; reproduced by permission of Chetham's Library.

Figure 2.17. Winchester Castle Great Hall, looking west; built c.1222–1235; photograph reproduced by permission of the Conway Photographic Library, the Courtauld Institute.

Figure 2.18. Winchester Round Table, built late thirteenth century; repainted c.1521; reproduced by permission of the Hampshire County Council.

Chapter Three

Of Sea Monsters, Salt, and Sovereignty

Edward I and the Gough Map of Britain

"THE IMPETUS BEHIND GEOGRAPHICAL MAPPING," HARLEY AND Woodward have said, is "usually the desire for territorial expansion and control." But military and administrative ambition tends to outrun its effective imposition or practice.[1] Through an examination first of its roads, then of some less strictly "geographical" features, this chapter will illustrate how the Gough Map of Britain instantiates both the above claims, with special regard to the participation of English royal clerks in Edward I's attempts to conquer and administer Wales and Scotland during the late thirteenth century. It will also argue that the Gough Map serves as a metaphor for a nascent English imperial state, whose asserted "monarchy of the whole island" is embodied through certain features included on the map which are likewise constitutive of the royal "fisc," a sphere of the public domain for which the crown acts as custodian and guarantor. In pursuing these lines of argument, I will challenge prevailing interpretations which read Gough as preeminently "a map for travellers," underscoring instead the function of English maps even in pre-Tudor times as tools of secular government which operate in modes both practical and ideological, through a cartography at once "factual geographic" and "symbolic."[2] What will emerge, in short, is a picture of Gough as essentially a document of colonial administration and propaganda, a map not suited simply to wayfaring generally, but one designed above all to engage the imagination of early English *imperial* travelers—in particular those aristocratic and clerical elites most invested in Edward's project of asserting overlordship in Britain.

Much caution has been used in differentiating the predominantly ecclesiastical cartography of the Middle Ages, overall, from that produced both beforehand, during Roman times, and subsequently, particularly in the age of European expansion. As the following material will show, however, it is not too much to say that the Gough Map of Britain is, like many later explicitly colonial maps, for all intents and purposes an administrative image made by and for the agents of empire. To demonstrate this we shall be discussing, first, the Gough Map's curious road system, and following this a selection of the map's other major iconic features, including its towns, its forests, and its rivers. This will lead to analysis of a few of Gough's more unusual features, including two brief written legends—treating, respectively, the salt of Droitwich and the landing of Brutus the Trojan off the coast of Devon. Finally, we will consider three strange drawings which appear in the Gough Map's North Sea—in the context, in particular, of what these illustrations may have to say about English relations with Scotland at the end of the thirteenth century. Overall we will find that Gough is a document even more remarkable than has yet been acknowledged—a medieval map that builds on the visual, cultural, and epistemological traditions activated by ecclesiastical *mappaemundi*, to be sure, but one that goes far beyond what medieval secular maps have been presumed capable of, in its representation of a complex yet specifically oriented political position.

ROAD BUILDING, REALM BUILDING

The Gough Map's road-system has been described as "extensive," but this signature feature of the document has a decidedly unfinished aspect to it. Gough's roads do not extend at all into Scotland or much into Wales, beyond one in the south out from London via Brecon to St. David's, and another along the northwest coast from Chester to Cardigan. Nor do its roads run in England anywhere north of Hadrian's Wall, on the well-traveled (especially following 1290) Scottish march (fig. 3.1).[3] For that matter, even in south England and the midlands the map omits a number of highways "obviously familiar" to the document's compiler, such as the route from London to Canterbury to Dover so important to Matthew Paris's cartography, not to mention to the Roman road system long before him and to writers on geographical themes later, including Geoffrey Chaucer.[4]

However, "many if not most" of the towns and villages on the Gough Map (especially the more obscure ones) appear to have been included simply because they lay along an important long-distance route. Sometimes these routes are marked with an actual line, sometimes not. Between London and Dover there may be no line for a road, in other words, yet the map sets out in proper succession the Channel-ward queue of Dartford, Gravesend, Rochester, Sittingbourne, Faversham, Canterbury, and Dover (fig. 3.2). Gough's land-travel system is thus at once much fuller than it seems (in that it textualizes numerous routes silently as it were, by way of an aligned "thoroughfare" of towns) and yet decidedly less full than it might be (in terms of its actual lines marking highways).[5]

The accepted explanation for this idiosyncrasy is that the Gough Map's compiler tended to draw a line representing a road only when he felt he knew the distance between the towns or staging points on a given route.[6] One result of such practice is that it assigns to England south of Hadrian's Wall, and to certain parts of Wales, a markedly different legal and administrative character than that imparted to the rest of the island, which is by implication wild and trackless, hence lawless. Geoffrey of Monmouth's King Belinus built his roads, after all, rather to establish the bounds of his political jurisdiction more firmly than to move around his goods and people more efficiently.

That no highways are drawn in Scotland and only two in Wales should not be taken to indicate a lack of interest in these regions' routes and itineraries, especially military ones. Evidence of very particular such interest appears in the Gough Map's road from London to St. David's. This road crosses into Wales just past Hereford (literally, the "army's-way," and historically a launching point for campaigns into Wales), then passes through Brecon and onward in eight stages to the coast (fig. 3.3). But in its western section this route follows a course which differs markedly from the main highway in this region, at least as it is known from other early sources and as it has run since the sixteenth century. In lieu of the main route (which like some forty percent of Gough's roads follows a Roman highway), the line entered on the map marks instead a mountain road which was as obscure then, apparently, as it is now. However, Gough's indirect route documents precisely a path that was once taken by Edward I, in the course of an expedition to put down Welsh rebellion in 1295.[7]

Similarly connected with the English subjugation of Wales (1277–95) are the balance of the Gough Map's sites in the north. The most unambiguous trace of this may be found in a series of castles newly built or rebuilt in these years in the neighborhood of Snowdonia, within whose mountain fastnesses Llywelyn ap Gruffudd's principality had its base. In general the Gough Map pays little attention to mountains and other high ground, but *Snowdonne* is given much prominence and drawn as a large scalloped hill, "resembling half a cogged wheel."[8] The Anglo-Norman/English castle has been described as the "symbol of domination *par excellence*"; but it is also, as the Gough Map shows King Edward well knew, conquest's most effective tool. Fortresses on an elaborate scale were thus erected by the English with considerable "speed and zest" not to mention expense, in a kind of coastal net around Snowdonia, for example at Caernarvon (Gough has *kaernarvan*), Conway (*conw . . .*), Criccieth (*crykkey*) and Harlech (*Harlech*), all beginning in 1277, and at Beaumaris (*beaumorres*) on Anglesey in 1295. The building of new castles was undertaken elsewhere as well, for example at Flint, Rhuddlan, and Denbigh further east—all sites which the Gough Map includes. Taken overall, it seems telling that a great majority of the places recorded by the map in North Wales (*Wallia borealis*) play a noteworthy role in the events of

1277 to 1295, either as focal points of Welsh rebellion (e.g., Hawarden) or as sites of English military and administrative activity.[9]

In a unique detail, the Gough Map records a measurement of the island of Anglesey (just off the north coast of Snowdonia) in the same Roman figures it uses elsewhere for road-distances (fig. 3.3, above). In three lines at the top (east) of the island is written *Insula de Anglesey habens xxiiij mil. in long. et xviij in lat.* ("the island of Anglesey has twenty-four miles in longitude and eighteen in latitude"). "So productive" according to Giraldus Cambriensis "that it could supply the whole of Wales with corn," rich-soiled Anglesey saw its crops, essential to the Welsh forces in Snowdonia, duly harvested by the English during occupation of the island in 1277, 1282/83, and 1295. On each occasion the island served the English as a convenient naval base (troops and supplies came on ships from the Cinque Ports, via Bristol). In 1282/83, moreover, Edward I's campaign plan hinged on employment of Anglesey as a staging point for invasion of Snowdonia; English forces were to cross to the mainland (on a ship-bridge across the Menai strait to Bangor) at the same time that the king attacked by land from Conway. Any of these periods of occupation would have provided an English surveyor or cartographer (perhaps a clerk in the royal household, as the king himself visited the island in 1283 and 1295) with the opportunity to measure the island, presumably after the manner used to figure the distance of Gough's 150 or so measured road segments.[10]

It is not unlikely that whoever measured Anglesey also played a part in the design, construction, or survey in these years of a new road linking Chester to Cardigan, via Bangor.[11] Built primarily in 1283/84, though incorporating forest cuts from 1277 as well as Roman work in some sections, this route progresses first across the north, past the king's new castles at Flint, Rhuddlan, and Conway. It curves southward at Caernarvon, whose elaborate castle was the most consciously imperial in its splendor of all the king's works in Wales, to run along the west coast by way of Criccieth, Harlech, and Aberystwyth, each likewise the site of a royal castle recently built or rebuilt (see fig. 3.3, above). Like the first, this second Cambrian road inscribes onto Gough's *Wallia borealis* a specifically royal itinerary—Edward I passed virtually the whole of this way in the course of a military perambulation he undertook to mark his conquest in 1284, and putting down rebellion in 1295 he again traversed the northern portion of it. 1284's royal progress from Flint to Cardigan was crucial in settling the details of the Crown's new administration: along the way Edward appointed permanent officials and established free boroughs, for example at Flint, Rhuddlan, Conway, Caernarvon, Criccieth, Harlech, and Bere ("the population of such new places being largely English"). But this conqueror's passage in arms along a newly established road also appears to have served, simultaneously, as a kind of processional adjunct to the king's industrious Round Table pageantry, held with Edward in the role of *Arturus redivinus* that same year at Nefyn in Caernarvonshire.[12]

Above we noted the role of raised highways during the imperial Roman occupation of Britain, as both practical tools and monumental markers of dominion. It is important, however, to be cognizant of the fundamental material difference between Roman-era highways and post-Roman trackways, the latter of which typically "grew from habitual lines of travel" (they "made and maintained themselves") whereas the former appeared as a result of centralized planning, implemented at an institutional level. As B. P. Hindle has stressed, "virtually no new roads were built" in the Middle Ages and of those that did appear, extremely few were "formally constructed or engineered."[13] Given the Gough Map's documentation not only of the course of a planned new route across Wales but moreover of the distances between its fortified stages (e.g., Cardigan *XXIIII* Aberystwyth; Harlech *XII* Criccieth *XXIIII* Caernarvon; Rhuddlan *X* Flint *X* Chester), it would appear that Edward's material alteration of the Welsh landscape, to both practical and symbolic ends, constitutes an important exception to standard medieval roadmaking practice, and one apparently conscious of its imperial precedent. The Gough Map connects Cardigan to Chester in a military chain each of whose links connotes English conquest. Several sites carry ancient pedigree as well—Caernarvon Castle with its walls recalling Constantinople was built beside the site of the Roman fort *Segontium*, and Chester was an important Roman fortress and highway hub, serving (like Hereford) as launching point for westward campaigning. In short, Gough's road from royal castle to royal castle around the wheel of Snowdonia describes an English conqueror's secure rein upon Wales, and reign over it.

Even the unit of measure the Gough Map employs underlines the military and imperial orientation of its route-marking project. This is because Gough's distance-figures, as the Anglesey-legend's use of the term *mil.* helps make clear, correspond to the Roman lineal measure of 1,000 paces (*mille passus* or *passuum*).[14] There is no direct etymological link between the unit of measure deriving from this, the Middle-English *mile* (c.1290) or *myle* (c.1300), and its near-homophone, the Latin *miles* ("soldier," "foot-soldier"; "knight"). However, given that the Latin *mille* ("a thousand") frequently operates elliptically, to signify "a thousand soldiers" just as it does "a thousand paces" (both are common usage), a certain slippage between the ME *myle* or *mile* as unit of measure and the military paces of the (singular or collective) Roman *miles* or foot-soldier seems not unlikely.[15] Such semantic play is particularly appropriate given the extensive overlap (both technically and historically) between measured routes, factual maps, and military campaigning. The measured paces taken by whoever gathered the distance-data for the Gough Map's Chester-Cardigan road were likely stepped off as part of or in preparation for a royal progress doubling as military circuit. Thus the conqueror's *milia milia* (or without ellipses, his *milia militis milia passuum*), his thousands of soldiers' thousands of paces, inscribe upon heretofore-trackless but now-subject Welsh territory the documentable administrative authority of an at-once imperial and vernacular English *myle.*

It has been asserted regarding Wales on the Gough Map that "the absence of roads proves nothing." In the spirit of some productive recent readings of absence in medieval texts, however, I contend that Gough's lack of roads (or more precisely, the absence of lines marking measured routes) across much of Wales, plus altogether in Scotland and on the Scottish March, indicates something more than, say, an arbitrary if regrettable documentary breach, or simply disinterest on the part of the map's compiler. Rather, this absence may bear witness to the relative unsettledness or unmeasurability (from a Westminster standpoint) of such provincial routes, as well as of the political territory they ought (on an "ideal" level) to penetrate, cross, structure, and define. Scientific techniques such as triangulation did not come into systematic use in the mapping of Britain until the mid-sixteenth century, but cartographic historians have long speculated that "a rudimentary survey" of some kind may underlie the Gough Map, considering its comprehensiveness and remarkable accuracy.[16] Given that a date of c.1280 has been suggested for the Gough Map's prototype, it may be that such a project of land survey was conducted as part of an extensive royal survey of another kind: Edward I's massive "Little Domesday" inquest of 1279–80.

Little Domesday was a comprehensive survey of tenements and liberties (or rights and properties) in England, from those of the nobility down to those of the peasantry, undertaken "in order that the king and others may know what is and ought to be his and theirs." This "immense enquiry" into medieval landholding is rivaled in scope and detail only by William the Conqueror's original Domesday Survey of 1086/87. Indeed, Little Domesday would have "utterly dwarfed" its predecessor, as R. H. Hilton has commented, had its commissions all been executed and had all its returns survived. Judging from those extant (commonly known as the Hundred Rolls), Powicke has suggested that once completed this survey would have provided "a combined gazetteer, terrier, and at least rural census for the whole of the country."[17]

In March 1279 Edward appointed panels of commissioners to "go personally to all and singular places" in every English county, "so that all the towns, hamlets, and other tenures shall be written in books, to be delivered to the king by the commissioners, in such a way that no one is spared, and everything relating to the premises is made fully clear to the king and others." However, returns survive only for some core midland counties, full ones for Oxfordshire, Cambridgeshire, and Huntingdonshire and partial ones for Bedfordshire and Warwickshire.[18] Perhaps the key point regarding Little Domesday's possible relationship to the Gough Map is that despite Edward I's successful first campaign against Llywelyn, affairs in Wales were in 1279/80 too unsettled to allow for effective inquiry by royal commissioners into the principality's landholding, even of the Crown's holdings there. In fact, the survey of Wales was not even attempted under this commission, however ambitious its contours otherwise.

With so small a sample, it is debatable whether much can be made of an absence of returns covering the southwest of England and the Scottish March. Yet it is precisely these regions that have been identified (along with Scotland and Wales) as road-thin and sparse of towns on the Gough Map, their geography least accurately depicted. Wales's roads are relatively few, as we have seen, while Scotland is especially poorly shaped and completely trackless—of course, both lay beyond the circumspection of Little Domesday's royal commissioners.[19] Taken all together, then, the status of England's administrative and political relations *vis-à-vis* Wales and Scotland at the time of Little Domesday provides a possible explanation for the Gough Map's notable failure to include routes within Scotland and the Welsh interior as a part of an overall British system. The map's compiler had arrived, critics have said, at the "conception of a national road system, based in London."[20] But although an England-centered *totius insulae monarchia* or "monarchy of the whole island" may have been envisioned, along the lines of Geoffrey of Monmouth's seductive Arthurian territoriality, nonetheless the full articulation of such an imperial system of measured routes covering the whole island (e.g., Totnes to Caithness) was not yet a practical possibility on the ground, for Westminster officials or their road-pacing, town-and-tenure-recording agents.

Gough's "national road system," to the extent that it sketches a geometry which is centered and radial, displays an underlying schematic structure which marks it as formally hence ideologically akin to other medieval maps, as noted in an earlier section. Yet this system does not depend in its geographical matter upon the legendary Four Ways of Britain, those roads which structure chronicle geography as well as Matthew Paris's *Scema Britannie*: neither Icknield Street nor Fosse Way appears at all on the Gough Map, and only a short length of Watling Street does.[21] In favor of such paradigmatic routes, when drawing lines upon Britain the Gough Map's compiler chooses to rely rather upon such highways as he or an agent might "go personally" upon, perhaps in fulfilling a royal commission as per Little Domesday: "to view all cities, boroughs and market towns," recording their size, status, location, and perhaps a measurement of the distance between them, in books "to be delivered to the king," all of it to be done "in such a manner that their [i.e., the commissioners'] circumspection may be perpetually commended, and so that they may not be confuted by a fresh inquisition or attaint for negligence or insufficiency." Rather than crib itinerary and other data from older written texts, as other medieval maps do, Gough relies instead upon the kind of comprehensive detailed information which can only have been gathered empirically, in the field, town by town, *mille passus* by *mille passus*. Just as the development of Vipsanius Agrippa's map of the world and of the first comprehensive portolan charts had required that data be gathered by specialist personnel and compiled by some central authority, given its similarly great geographic ambition the Gough Map, in the course of its development, must have required and been accorded extensive institutional support—the likes of which, it appears, England saw activated in 1279/80's Little Domesday, "the greatest enquiry of its kind" since the Norman Conquest.[22]

Given that the Gough Map's topography, town-distribution, stage-to-stage distance-measurements, and even incidental marginalia were copied faithfully by English cartographers through into the middle of the sixteenth century, it would seem that its makers met the challenge of perpetual commendment set by Little Domesday's commission. Once drawn together by the king's administrators, this map didn't need to be recompiled for some two-and-a-half centuries—and indeed seems to have been definitively updated only with great difficulty.[23] If Gough's roads do not pass into Scotland or (beyond the royal itineraries noted above) into Wales, perhaps this is because Little Domesday's commissioners didn't either.

In initiating Little Domesday, Edward cited a "desire to redress the state of his realm." In practice, this translated into an attempt to "reestablish wherever possible royal rights which had lapsed in earlier years of weak monarchs." Far from disinterested, Edward's "great visitation" was thus another of his reign's "manifestations of royal power," an act whose institutional expression came at direct expense of all landholders below the Crown.[24] To put it another way, the king's metropolitan conception of the space of his kingdom (i.e., as centralized, bureaucratic, and fiscal) was in this project of territorial demarcation and administrative enrollment asserted against those models of geopolitical organization to which other social groups subscribed. Little Domesday lay siege in various degrees, that is, to the cosmopolitanism of mercantile interests; to the regionalism of certain baronial groupings; to the localism of an increasingly shire-based lesser gentry; and to the village communitarianism or parochialism of the smallest landholders and those tied to the land. The losses from encroachment sustained by "both rich and poor" were to be amended by the king's inquest, but the balance of those subjected to "faithful examination" by Little Domesday's commissioners probably were inconvenienced by and much resented such a deep incursion of royal power into local affairs, meaning the king's surveyors were in all likelihood no more happily suffered than were his military purveyors, itinerant justices, and revenue collectors.[25]

In pursuing such exhaustive documentation, not unlike a conqueror's enrollment of all the land's tenures, fees, and liberties down to the last sub-holding, the royal administration asserted the King of England's proper responsibility as guarantor for all of this, asserted Edward I's ultimate jurisdiction, as custodian of the public good, over the whole of the realm and all its resources. Thus measured were the bounds, Kantorowicz or Bracton might say, of the impersonal, ubiquitous, and inalienable royal fisc. The king's array of custodial rights as documented by Little Domesday comprised thus a kind of public sphere which pertained to all, which is as much as to say that this commission surveyed the royal fisc, that embodiment of the public domain which came in precisely this period to represent both state and prince, and to serve as sovereignty's hallmark.[26]

Historians have long regarded the written texts which register legal and administrative action as preeminent among the markers of conquest: William I's Domesday Survey expresses Norman domination far beyond what the totemic Battle of Hastings achieved itself, with the Bayeux Tapestry coming perhaps a distant third in terms of such service.[27] But because it incorporates both these modes (the administrative and the pictorial), the Gough Map of Britain produces the effect both of survey and of tapestry; its military road across north and west Wales, from royal castle to royal castle around the wheel of Snowdonia, both commemorates Edward's conquest and ties this territory into the measured bureaucratic net of English government, as enacted in 1284's Statute of Rhuddlan. The conquest of Wales came too late to see it comprehensively surveyed under Little Domesday. But a measured road or two tying a string of new boroughs in Wales back to towns just across the border in England, i.e., Hereford and Chester (fig. 3.3, above), might signal even so an incorporation under the governance of Westminster of the Crown's newly conquered territory. It is from Hereford and Chester, incidentally, that templates were borrowed, in producing the town-charters for Wales's new boroughs.[28]

Scholars concur that Edward I's conquest of Wales was an achievement more to be put down to effective organization ("the massive mobilization of men and resources") than to "any brilliant stroke of generalship." Such was also the case with England's incursions into Scotland. In both these military theaters, the actors performing the "essential work" were as much the king's administrators as his soldiers.[29] Edward's Welsh and Scottish wars marked all of English society deeply, in particular from about 1295, after which a state of war was more or less constant and ever-increasing numbers were called on to participate in the fighting or to support it materially. This lent the enterprise "something of the color of a national struggle": "it was a situation where *quod omnes tangit, ab omnibus approbetur* ["what touches all, all must approve"] and where all were called upon to defend the land against a common enemy."[30]

From the standpoint of administration, English conquest of Britain was above all a project to be orchestrated by clerks of the royal household—although to be sure this was an activity both more strenuous and more multifaceted in its demands than most. In later years, royal officials who had been active under Edward I looked back to this "as the one understandable period in their official careers"—a time "when plans could be made effective and things could be done." The range of "things done" by English royal clerks under Edward I appears to have been extremely wide, with duties for the most part unspecialized.[31] But just as it is officials like these who commissioned and compiled Edward's 1279 Little Domesday inquest, so, similarly, are king's clerks precisely the class most likely to have been involved in the conception and compilation of a Gough Map prototype and in its subsequent recopying. This would be true if for no other reason than that in royal clerks we see combined a multifarious array of talents and duties, together with a concern for the community of the realm (*communitas regni Anglie*) which during these years bent, above all, to the project of conquest.[32] But that those very officials who organized Edward's British wars may have had cartographic inclination (unusual by the standards of a comparatively map-thin culture) is suggested in the chance (ranging from possibility to likelihood) that the Gough Map's first known owner acquired it through a government source. Less in question but arguably as important is the full integration, as seen during the sixteenth century, of territorial maps into the administration of the realm—documents initially based on though eventually moving away from Gough's model. Overall, it is surely noteworthy that those officials who played a role in surveying the realm and in organizing conquest also had an interest in cartography.[33]

We said at the outset of this chapter that administrative ambition—whether bureaucratic theory or military aspiration—will tend to outrun its practice. In its incomplete network of roads, the Gough Map thus appears to concede that there are limits to English administrative authority in Britain. However, in features other than its measured road-system the Gough Map asserts just the opposite: that properly Britain is a single monarchy of the whole island, unified under the overlordship of the king of England, heir of Brutus, Belinus, and Arthur.

Just above we invoked the concept of the royal fisc, a sphere of public good (including both rights and resources) more or less concomitant, theoretically, with the communal body of the king. As opposed to his physical body or "body natural," the king's public body never dies, and is itself coincident with the territorial bounds of his kingdom; furthermore its rights, properties, and possessions are, like those of the church, inalienable. The Gough Map embodies one aspect of this concept in the signature feature of its traffic-bearing but moreover jurisdiction-carrying royal and public highways. Yet as Kantorowicz rehearsed for us in chapter two, also central to the royal fisc, beyond roads to carry peace and justice, were various other "things," properties or institutions pertaining to the public good (*res publicae*) such as rivers, forests, and public squares.[34] These are also features which the Gough Map highlights, as the following pages will detail.

RIVER, FOREST, TOWN, SALTPIT: ICONS OF THE STATE

Rivers are "numerous and conspicuous" on the Gough Map, as on many other medieval maps. This may be because waterways were at least as important as roadways to travel in the period—in addition to being safer and more reliable. Rates of town-growth in medieval England link closely to the factor of navigable water. And indeed roads and tracks often served primarily "as feeders to the river system" (especially in the case of heavy goods or important personages), a practice which underlines the overall integration of the realm's waterways (including coastal routes) with its land travel system.[35]

Given such factors, the pertinence of rivers to the public good, and hence their place among the elements comprising the royal fisc, need come as no surprise. Accordingly, just as for the realm's "four public roads" or King's Highways, the Laws of St. Edward the Confessor provide for the protection of "a higher jurisdiction" upon the realm's "principal rivers" (those "known by name" and by which "provisions are carried from different places to cities or boroughs"), as well as for their maintenance. Provision for the "lesser rivers" is also made—exactly "as prescribed by the law relative to the lesser roads."[36] In this light the Gough Map's attention to rivers seems to constitute something beyond just the usual practice in medieval terrestrial cartography, and appears to serve a function beyond merely that of better enabling the movement of goods and people, to which end the map's remarkably close river-knowledge has heretofore been attributed. In addition to this, Gough's careful enrollment and demarcation of all the realm's waterways, major and minor, may signify its accounting of this component aspect of the king's fiscal rights and responsibilities, an articulation of the contours of the public domain as against the realm's various private landholding interests and privileges. In this respect, Gough operates as an administrative survey after the general nature of 1279's Little Domesday, one not only in line with the project's underlying rationale but indeed its specific details: among other tenures, the commissioners were to enquire into "waters and rivers."[37]

Themselves and alone, it may be that the Gough Map's roads and rivers do not significantly undercut the proposition that this document is first and foremost "a map for travellers"—rather the opposite, since measured roads and accurately drawn rivers are precisely the features cited in support of such interpretations. Certain other features of the map fit less well with such a characterization; yet these features—those "a little unusual for a map meant for use by travellers"—are exactly those which support the proposition that the Gough Map serves in part to represent the royal fisc.[38]

In this regard the legend *hic fit sal* ("salt is made here") beside the town of Droitwich in Worcestershire stands out. Possessor of a modest icon and only half-legible name (see fig. 3.4), this site would hardly distinguish itself from the rest of Gough's hundreds of settlements—except for its unusual characterization through even such a laconic inscription as three short words. For this map has only a few such pendant-legends, particularly in England, and indeed very little text overall—Gough is *not* a map "at least as much written as drawn" after the manner of many *mappaemundi* or even of Matthew Paris's Apulia or *Scema Britannie*.[39] This being so, the Gough Map's curious gloss beside Droitwich seems to token a concern on its compiler's part worth our noting, for matters beyond simply the more effective planning of travel.

This is not to say that Droitwich's notation as a center of salt-production is irrelevant as regards medieval wayfaring. Routes to the brine pits at Droitwich, as to the salt-mines of Chester, appear to have been extremely numerous in the period. The Gough Map eschews representation of such local "saltways" (as described and diagramed by Hilton), but it does mark two long-distance roads into Droitwich. Neither of these forms a leg or branch of one of the map's major radial highways from London, and indeed one of the two begins here—a point which bespeaks the town's status as a destination in its own right, rather than as merely a way-station. Given the road and river integration noted above, also noteworthy is the inclusion of the River Salwarpe (*fluvius salw . . .*), a navigable tributary of the Severn flowing between Droitwich and Bromsgrove.[40] Droitwich serves in short as a minor hub in the Gough Map's travel system.

The mode of utility which the Gough Map's *hic fit sal* legend at Droitwich introduces appears to be based in a concept of public interest, one relating to the king and his clerks insofar as relating to the community of the realm as a whole (rather than simply the prosaic wayfaring needs of the crown's various itinerant servants). Salt, of course, was "vital to the medieval economy," primarily as a food preservative (e.g., for meat, fish, cheese, butter) but also for industrial uses, such as tanning. Private individuals (typically the lords of local manors) owned lucrative rights to shares in the brine, distributed by vat then boiled down into salt using fuel from local woodlands.[41] However, a certain degree of public or royal interest in salt and hence saltways may be inferred on the basis of contemporary military purveyance accounts. For use on Edward I's Scottish campaign of 1303, for example, more than 680 quarters of salt were ordered (and duly enrolled upon receipt at Berwick), in addition to salted meat and fish including 447 carcasses of pork alone.[42] But beyond what might well represent an inordinate royal concern for the product of salt as such—wartime victualling needs were not to be denied—it may be that king's clerks took on a custodial interest in this resource and the roads thereto as a component of their duty to safeguard the greater public interest, to maintain the collective rights of the royal fisc. Hindle has said that "saltways were certainly not special roads in any sense," except insofar as place-names like Salterford make them relatively easy to trace. However, the Crown's duty in seeing them maintained in the public interest, as utilities pertaining to all, if especially relevant to military purveyance, helps suggest how the royal administration's attention to travel-routes and their maintenance may go beyond simply the question of efficient routing of people and goods.[43]

We begin to touch here on the question of a centralized bureaucracy's relationship to national, regional, and even local economic affairs, and its stewardship of production and commerce, of the realm's resource extraction and distribution, so as to benefit the collective good (conceived in terms of the Crown), not just enable the profit of individuals.[44] There is little evidence that English royal clerks in the thirteenth century ever thought to formulate anything resembling an explicit national economic policy, and then direct or implement it, after the manner of Frederick II's officials in Sicily. Still, if any species of such consideration took place as the words *hic fit sal* were added to the Gough Map, whether in the mind of this map's original compiler or some later redactor or glossator, we may be said here to witness "use" on a different level altogether from that inherent in the accepted image of some king's courier, letter in hand and horse waiting outside, peering at a copy of the Gough Map hung at Westminster and plotting out, then committing to memory, the stages in his quickest route to Worcester.

The Gough Map's gloss characterizing Droitwich in terms of its primary economic product, salt, recalls the practice of a fascinating literary text composed c.1290–1330 in a mixture of Anglo-Norman and Middle English. Known to scholars as the *List of 108 English Towns*, this curious survey or geographic poem consists of an at times obscure selection of Britain's settlements and districts each paired with its staple product or signature attribute: e.g., *Samon de Berwik* ("Salmon of Berwick," line 37), *Corde de Bredeport* ("Cord of Bridport" 90), *Payn de Seynt Alban* ("Bread of St. Albans" 69), *Archiers de Wals* ("Archers of Wales" 47), *Nauie de Suthamton* ("Navy of Southampton" 60), *Corn de Cardoyl* ("Corn of Carlisle" 78), *Escole de Oxenford* ("School of Oxford" 9), and so forth.

In one of just a handful of notices, this text has been accorded the memorable judgment of having been composed "by a clerk in his cups," an impression which undoubtedly stems from its disconcerting habit, in pairing each place with a single attribute, of drifting between the concrete and the abstract.[45] In the *List*'s opening sequence, for example, after beginning with the *Barounie de Loundres* ("Barony of London") we jump from the *Seyntuarie de Canterburg* ("Sanctuary of Canterbury") and *Relikes de Westmoster* ("Relics of Westminster") to, jarringly, the *Puteynes de Cherring* ("Puddings of Charing Cross"), then from the *Pardoun de Seynt Pol* ("Pardon of Saint Paul's") to the *Sause de Flete* ("Sauce of Fleet") (1, 3–7).

As begins to become clear, the sheer range of the *List*'s subject matter is stunning, especially for a work so short (fewer than 350 words all told) and so circumscribed syntactically. One realm treated is the economic, at the level either of a town's staple product, such as the *Morue de Grimesby* or "Cod of Grimsby" (74), or of its characteristic profession, for example the *Poter de Henneham* or "Potter of Henham" (52). Also appearing are topographical attributes, whether natural landscapes like the *Pleynes de Salisbury* or "Plains of Salisbury" (26) or on the other hand feudal divisions such as the *Maner de Wodestok* or "Manor of Woodstock" (100). Similarly, religious characteristics appear not only at the level of institution, as for example in the *Encloystre de Lycheffeld* or "Cloister of Lichfield" (27), but also at the level of social practice, as in the *Pelerryn de Schrowesbury* or "Pilgrims of Shrewsbury" (45). Several attributes are military, as noted above in Wales's *Archiers* and Southampton's *Nauie*, as well as in the *Bachelerie de Norhampton* or "Bachelery of Northampton" (23), and in the *Marche de Punfreyt* or "March of Pontefract" (93). A few are legal, such as the *Playz de Wychelsee* or "Pleas of Winchelsea" (32). Finally, a large number of entries are less easily classifiable in institutional terms, such as local reputations including the *Dames de Seynt Edmo* or "Dames of Bury St. Edmund's" (8), qualities like the *Orgoyl de Bourk* or "Pride of Peterborough" (103), food preparations such as the *Furmage de Gerwaus* or "Cheese of Garway (?)" (82), and more.

All maps, as social geographer David Harvey once remarked, contain a measure of "pure poetry." So too this short text of social and geographical accounting. Perhaps the greatest temptation, given the *List*'s many engaging details, is simply to designate this text a figure (as indeed I have done; see fig. 3.5), and reproduce it in full for the poetry of its local color. To stop here, however, would be to fall in line with the prevailing critical habit, which has been to dismiss or disparage the work's invocation of survey format (because it appears to lack an "overall geographic coherence") and to regard the *List of 108 English Towns* instead merely as a trove of idiosyncratic detail, no more than social bric-a-brac.[46]

In fact, this survey-text depends very heavily on its geographical structure, departing from its three-word enrollment formula ("attribution *de* place") only in its concluding four lines. One result of this compositional discipline is that it enables the *List of Towns* to construct, sometimes along topographical lines, sometimes in thematic progression, a vision of Britain as an at once unified (politically) and diversified (e.g., economically) single realm. The net effect of such a litany of place-names and their characteristics is to introduce, into contemporary discourse on the political construct of the English feudal realm *cum* bureaucratic imperial state, a level or mode of administrative accounting far removed from that aspired to, much less approached, in even the most purportedly comprehensive of officeholding or landholding surveys.

The *List*'s parameters, in other words, are not so narrowly circumscribed as the *Notitia Dignitatum*'s or Little Domesday's, though its compilation of material is of course far, far less exhaustive. The mosaic this text assembles of medieval society is, if thinner, far more ambitious in color and scope; and if uneven, it is more far-reaching in its apprehension of medieval English society as a mix of factors, topographic and economic certainly, but also social and cultural. Through its irresistible accrual, in short, the *List of 108 English Towns* manages to invoke for its reader an almost tangible conception of the island of Britain as multilayered social territory, a body of collective human terrain comprehensible not through narrowly institutional or topographical description, but only through enrollment of data on a multitude of levels, including—perhaps most importantly—something of the poetry of local daily life.

Despite appearances to the contrary, the *List*'s selection and ordering of its towns and attributions is not completely arbitrary. In places we progress after the manner of other textual surveys, with sites entered according to their general geographic proximity: thus Winchelsea and Rye inhabit consecutive lines (32–33), for example, and as noted above some half-dozen sites in the environs of the metropolis appear in the work's opening lines, plus another (King's Cross) in line 16. Elsewhere, however, the *List* falls prey to its project's associative possibilities and all sense of a strict geographical progression (along the lines of the *Notitia Dignitatum*, say) breaks down. Yet in place of regional clustering there appears, frequently, another kind of coherence, an ordering principle based in the character of a site's attribution rather than in its geographical location. Thus we move from the *couert* ("lodging, shelter") of Sherwood, where Edward I had a hunting lodge, to the *chace* or private hunting ground of Inglewood and then to the *forest* of Windsor (75–77); from the *larroun* or "robber" of Graham to the *murdresours* of King's Cross (15–16); from the *couerchef* ("kerchief, headcloth") of Shaftesbury to the *wymple* of Lewes (43–44); and so forth. Among the more intriguing sets are the *List*'s progression from the *trespas* or "transgression" of Chelmsford to the *symenels* or "simoniacs" of Wycombe (39–40); from the restorative *Bayn de Baa* ("Water of Bath") to the *Merueille de Stonhengh* ("Marvel of Stonehenge") (28–29); and from the *Hardement de Cinkpors* ("Hardihood of the Cinque Ports") to the comparable *Orgoyl de Bourk* ("Pride of Peterborough"), although this comes by way of the concrete *Chastel de Doure* or "Castle of Dover" (101–103).

Once a given thematic vein has run dry, the *List* tends to proceed by geographical proximity again, or by generic type (e.g., another port, marcher town, or monastery: *Waltham-Seynt Alban* 68–69), or else by linguistic similarity (e.g., shared suffix: *Gypeswyz-Doneswyz-Northwyz* 66–67, 70; *Neubury-Schaftesbury-Schrowesbury-Tillesbury* 42–43, 45–46; *Henneham-Notyngham-Eylesham* 52–54). Still, to be sure many jumps between sites defy explanation altogether—as the progression between items on any list will.

Perhaps more important than such line-to-line principles of progression as these, however, is that on close examination there emerges an observable pattern as to when the *List of 108 English Towns* selects an economic attribute for a site, whether product or profession, and when not. In his seminal study of the West Midlands during the reign of Edward I, Hilton has shown that this region's economy was "on the whole unspecialized," in spite of exceptions such as the production of salt at Droitwich and of iron wares at Gloucester, the marketing and manufacture of cord at Warwick, and the importing of wine at Bristol. In reading Hilton against the *List of Towns*, it becomes clear that those towns which are given an economic attribution by the *List* tend, similarly, to be those cited by Hilton as sites of specialization. On the other hand, those towns the *List* characterizes with another order of attribution (i.e., along non-economic lines) tend to emerge in Hilton's analysis as sites of economic non-specialization. Hence there appears to be little arbitrarity after all in the *List*'s choice to characterize Warwick by its cord (*Corde de Warwik* 55), Coventry by its soap (*Savoun de Couentre* 84), Bristol by its wine-trade (*Vend' de q'ts de Bristowe* 88) and Gloucester by its forges (*For de Gloucestre* 25), as opposed to providing feudal, topographical, religious, or miscellaneous social characteristics for these towns.[47]

The net result is a kind of radically streamlined social, economic, and cultural survey, which has similarities at one extreme with Domesday Book, from which Chester's *furur* or "furriers" are known, and at the other with the conventional place-name collocations of literature and popular culture, from whence for example the *rymeour* or "rhymers" of Worcester, the *demayseles* or "damsels" of Hereford, and perhaps, via the ballads of Robin Hood, Nottingham's *boues* ("bows").[48] Translated into visual form, the *List of 108 English Towns* might resemble most a cross of the medieval *mappamundi*—Elephant of India, Skier of Norway, Transgression of Eden, Labyrinth of Crete, Crucifixion of Jerusalem—with the modern schoolhouse map of specialized production: oil rig in Texas, lobster in Maine, wheatsheaf in Kansas, dairy cow in Wisconsin, automobile in Michigan. As such, the map compiled by the *List of 108 English Towns* might be said to illustrate all told something of the contours of a late thirteenth-century royal clerk's professional purview, the variety and scope of his knowledge and concerns, as extended across the geographical canvas of the realm.

A number of the *List*'s attributions are all but irrelevant, from a narrowly administrative standpoint. But a great many of the qualities included may be described as pertaining to the community of the realm, hence helping to comprise a public fiscal sphere. Some hold obvious relevance for royal officials—military attributions such as navy and bachelery (i.e., a levied company of lesser knights). But more broadly conceived, the royal clerk's mandate in these years of militarism was to procure all manner of resources for deployment in the king's wars—not just archers of Wales but cod of Grimsby to feed them, russet-cloth of Colchester to clothe them.[49] The *Relikes de Westmoster*, for that matter, recall the "conscious campaign" undertaken by Edward I and his officials to establish Westminster as the symbolic as well as administrative center of a unified Britain. Hence the physical attachment to St. Edward's shrine of the Crown of Arthur, sign of Wales's sovereignty, in 1284, and of the Stone of Scone, Scotland's coronation stone, in 1296.[50]

The Gough Map of Britain attests for the most part to a narrower, a more topographic and administrative geographic vision than does the *List*. But rather than dismiss the *List of 108 English Towns* as a formless curiosity, the product of alcohol and free association, and Droitwich's *hic fit sal* as a meaningless idiosyncrasy on a map for travelers, better to see in both these texts evidence of a close link in the minds of medieval civil servants between the technology of survey (whether textual or cartographic) and the Crown's array of territorial assets and custodial rights. It has been said that the king's clerk in the reign of Edward I "stands precisely at the point where policy metamorphoses into practice and thereby creates polity": "Going out among the nation, they [royal clerks] forge the links and bring about a fusion of interests between king and subjects." It may be, therefore, that what we are seeing in these documents (both apparently produced in this institutional context of an "unusually high . . . level of public service"), is essentially an assertion by such officials of their responsibility for stewardship of all the realm's assets, its various products, institutions, and cultural and material landscapes and resources: those myriad qualities which together comprise the royal and public fisc.[51]

The *List of 108 English Towns* includes a number of attributions relating to the realm's woodlands—the consecutive *Couert de Shirwode*, *Chace de Engelwode*, and *Forest de Wyndesore*, for example. Similarly, the Gough Map's concern to include social and economic over and above locational data (beyond simply in the case of Droitwich) may be seen in its treatment of Britain's forested land. In four places, the map's compiler has drawn an icon consisting of two intertwined trees. In one case, on the Isle of Bute to the west of Glasgow, this icon has no label (see fig. 3.6). But the others are marked: *foresta de dene* ("Forest of Dean") in Gloucestershire, *Nova Foresta* ("New Forest") in Hampshire, and *S . . . w . . .* ("Sherwood Forest") near Nottingham. A fifth such site, in Cumberland, has the appropriate label, *foresta de Ingelwode* ("Inglewood Forest"), but no intertwined trees; and a sixth, at a site now known as Gaick Forest in Scotland's *plaga de baunagh* ("District of Badenoch"), seems to be implied by way of the drawing of a deer and its accompanying legend: *Colgarth hic maxima venacio* ("Colgarth; great hunting here") (fig. 3.7).[52]

The absence on Gough of a twinned-tree icon to mark the great woodland of Arden in Warwickshire has occasioned some confusion, as to whether to regard the notation *Arderne*, which does appear (in a cartouche such as frames most of the map's regional names), as reference to a forest or a district. There would not likely have been such confusion in the thirteenth century, for Arden, though ancient and extensive, was not a royal forest; hence its lack of such an icon as marks Inglewood, Sherwood, Dean, and New Forest is categorically appropriate.[53] What this illustrates is that the map's conception of a forest is not primarily topographical. Arden may bear noting as a district (that is, administratively), but it does not merit any special mention or iconographic treatment as a natural phenomenon relevant to wayfaring.

In this respect the Gough Map may be said to treat wooded land in much the way it does high land—both are given little attention, surprising insofar as these features would seem to be highly relevant to a map for travelers, in comprising impediments or complications to wayfaring. But even Gough's notable lack of interest in mountains and hills (relatively speaking; there are exceptions, especially in Scotland) pales in comparison to its absolute disregard for woodland as topography *per se*. If getting around were the primary point of the map, more important than to mark administrative districts would be to register the location of impassable or dangerous terrain. This issue was far from a minor one: for military maneuvers against Snowdonia, the Crown was regularly obliged to employ large numbers of woodcutters, to carve paths for the king's troops, workers, and supply trains through the dense woods of North Wales. The campaign of 1282–83, for example, saw the impressment of some 1400 woodcutters from half-a-dozen English counties, and in August 1277 alone as many as 1800 were engaged in cutting a road to Rhuddlan. Following conquest in 1284, provisions were made to ensure the widening of all "public roads" through Welsh woodlands, with trees on either side to be cleared away to a fixed distance, not so much to quicken movement along them as to reduce the threat of highway robbery.[54] The principality's dense woodlands clearly had decided implications for English wayfaring; but not a single tree-icon appears in Wales.

The great woodland of Arden's lack of an icon to represent it on the Gough Map, then, underlines the point that medieval forests (both generally and those on this map) were far less a topographical phenomenon than a social one. As Hilton has said, "In the Middle Ages the word forest did not, as it does now, mean a large area of wooded country. It had primarily a juridical meaning. A forest was a region under the law of the forest." To be sure, forests provided resources like timber, iron, and coal, plus pasture and small game for nearby populations, and were moreover "honeycombed" with cultivated land, host to villages as well as individual peasant holdings.[55] But these were above all large-game preserves, a theater for the fetishized social performance of the aristocratic hunt and, as such, sites of particularly intense legal attention (as well as the avoidance of same). Gough's inclusion of half-a-dozen textually labeled and/or iconographically marked forests, then, attests not to the relative density here of trees, but rather to the special purchase or intensity in these regions of royal administrative and juridical authority.

There are a number of ways in which this representation of royal forests relates to the Gough Map's apparent project of delineating the bounds and contours of the royal fisc. Like other woodlands, a royal forest will pertain to this sphere of public utility insofar as embodying a variegated national economic resource—provider of timber, coal, arable land, and so forth. But more specially relevant is the royal forest's status as an inalienable possession of the Crown, whose rights, perhaps here more than anywhere, were not to be encroached upon. In fact, one of Edward I's handful of references to the nebulous (because oral, vernacular, and uncodified) fourth clause in his coronation oath, that engaging him to maintain unimpaired *l'estat de la coroune* ("the rights and territories of the Crown/kingdom") and to recover all those which had been lost, came appended to "certain proposals" he made at Parliament in 1301 "regarding the perambulation of forests, which he had reason to suppose might not be acceptable to the magnates."[56] His rights in this territory, in short, pertained to the king's very sovereignty. And not only, then, is the medieval forest, as evidenced by Gough, less a topographical than a social, economic, and legal category. What is more, to the extent that they are explicitly designated as falling under a special royal jurisdiction, forests, like the realm's roads and rivers, constitute another in Gough's accrual of the features of the ubiquitous extra-feudal sphere, another element in the map's illustration of the collective body of the royal fisc.

In the enthusiasm for discussion of roads and travel, commentary on Gough has shortchanged most of all interpretive iconographic assessment of the map's over 600 towns and villages. To the extent that towns in alignment comprise regional and long-distance thoroughfares, and hence assist in route planning, this major feature of the document may appear merely to reify Gough's traditional characterization as a map for travelers; great care is taken, after all, to see that each settlement vignette is placed in precisely its proper location, relative to coasts, rivers, roads, and other towns.[57] Read in terms of these icons themselves, however, Gough's towns argue a point directly opposite of that which its road-system seems to argue, with its acknowledgement—through an absence of actual lines marking measured routes in Scotland and much of Wales—of certain limits to Westminster's administrative sway.

Relatively little work has been done in historical cartography treating the various topographic conventions in use on medieval maps, a point especially true of town-icons. But one feature at least has been noted regarding the Gough Map's architectural vignettes: that these icons appear to indicate relative status. Typically, the size and general character of a town is indicated—cathedral cities and monasteries have crossed spires, and military sites have towers and/or crenellation. However, categories are not clearly demarcated—there being not so much a few easily separable types as a continuum incorporating "many combinations of symbols," from single buildings to elaborate walled towns containing spired churches, multiple towers, and numerous houses.[58]

Unlike other medieval maps whose cities are frequently drawn to include a measure of unique local detail, in rendering its town-icons the Gough Map avoids such individuating practice almost entirely. Instead, the differences between sites are expressed generically, through a differently combined but ultimately limited set of iconic building blocks (house, spire, tower, wall) rather than through any kind of truly documentary detail, such as characterizes Lincoln or Bethlehem on the Hereford Map, for example.[59] In other words, despite considerable variation from icon to icon in terms of the degree of elaboration and the particular combination of symbols, all Gough's settlements are made of the same architectural stuff, constructed from a few reusable iconic elements.

One result of this is that except in their relative density and size, towns in England do not differ identifiably from those in Wales, in Scotland, or on the Scottish march. Thus there prevails on the document an overall coherence of architectural style, a uniformity at the level of urban sign which implies a similar uniformity at the level of collective community identity, perhaps even at the level of political character or destiny. In short, through a lack of distinction between those in England and those in Scotland or Wales, the Gough Map's town-icons articulate the point that all the communities of this territory, all the inhabitants of this island-realm of Britain, combine to make a single, London/Westminster/*familia regis*-centered whole. Thus expressed is Geoffrey of Monmouth's basic territorial ideology, as well as King Edward I's. And just as with the rivers and roadways connecting them, all the island's cities, boroughs, towns, and hamlets, as well as the charters, markets, fairs, and liberties of each (so Little Domesday's commission), lay within the bounds, properly and historically speaking, of the fiscal authority of the king of a unified Britain.

This sense, however, of a social and political unity lent to the Gough Map's settlements in Britain by their overall iconographic coherence, does not extend beyond the island. Roughly fifteen percent of the territory included on the map lies beyond Britain and its coastal islands, thin stretches of *Hibernia* (at the bottom) and *Norway*, *Dacia* (Denmark), Flanders, and France (at the top) which provide a kind of schematic frame around Brutus's promised island and its waters. There is certainly space enough, but of the map's handful of town-sites in these territories (e.g., *Sklus* or Sluys in Flanders, *Develyn* or Dublin in *Hibernia*, and *Boleyne* or Bologne in France) only one is accompanied by an architectural vignette such as marks literally every town or village site noted in Britain. This exception is Calais, doubtless provided with an icon because following its siege and capture in 1347 it came to be regarded as an English town, serving as a key site both in war and in commerce.[60]

As Calais underlines, what is finally most compelling about the Gough Map's towns is the way in which the set of them, while serving in their careful placement as an aid to travelers, yet effectively transcend a wayfaring function. This point may be brought into focus through consideration of another work of premodern technical cartography, a c.1500 view of Venice made by Jacopo de' Barbari. Juergen Schulz has observed of this woodcut print that "the detailed modeling of the city's buildings gives the entire urban fabric a uniform gray tone that sets it off from the more distant zones and draws it together, into a compact unity." Because Venice's buildings are coherent in style, even while being documentary and individual in their detail, the view "imparts to the city fabric a homogeneity." The woodcut's geographic matter is exceedingly, even obsessively "accurate"—and for this reason it had been seen by cartographic historians prior to Schulz as preeminently "practical," a street map or guide for walkers of the city. However, the actual subject of this print, as Schulz comes to argue persuasively, is not Venice as physical city or architectural phenomenon, but Venice as political construct, the mercantile commonwealth that looms as Europe's pre-eminent maritime power: "Her physical features are exhibited as the material manifestation of the state," meaning ultimately "Jacopo's print is a visual metaphor" which conveys "abstract notions about the commonwealth."[61]

We have seen that the Gough Map is characterized on one hand by its lack of a measured road-system—which is to say an English royal administrative and legal presence—both in Scotland and on the Scottish March, to whose unsettledness in the period chronicler Peter Langtoft, for one, bears shrill witness.[62] This state of only a partial knowledge of the north on the part of Westminster/*familia regis* bureaucrats is further attested by the Gough Map's strikingly poor depiction of the outline of Scotland, relative to the rest of the island; by its "inadequate" and "erroneous" representation of Scottish inlets and rivers; and by its relative preference here for floating district-names (e.g., *comitatus de Boghane, plaga que dicit. Loren*; "Earldom of Buchan," "District of Lorne") over discretely sited towns or villages: the ratio of districts to settlements is about two-to-five in Scotland, but less than one-to-twenty in England and Wales. The Gough Map's, hence English clerks', comparatively "crude" knowledge of Scotland, finally, is evidenced by a much greater frequency here of legendary material, a reliance so to speak upon the geography of hearsay—for example, Loch Tay's *in isto lacu tria mirabilia insula natans pisces sine intestinis fretum sine vento* ("in this lake there are three wonders: a floating island, fish without intestines, and a strait without wind [i.e., a calm section of water]").[63]

Yet despite these limits to its knowledge, the Gough Map emerges as a document which asserts the King of England's rightful inalienable lordship over Scotland, just as over Wales; and as a document, moreover, representative of the multivalent royal fisc, the collective body of the king which never dies and which is ubiquitous in Britain, taking in the red deer of Colgarth as surely as the salt of Droitwich.

Like Jacopo de' Barbari's view of Venice and Vipsanius Agrippa's world map at Rome, the Gough Map of Britain is thus a document whose geographical matter appears, at first glance, to be empirical simply for its own or for a wayfaring sake, hence the standard assignment of "practical" function to a "factual" nature. However, like Jacopo's metaphor for an expansive Venetian maritime state and Agrippa's display of Augustan Empire's extension over all the known world, this map's meaning must ultimately be recognized as ideological. No less surely than explicitly symbolic maps, factual geographic maps operate on an "ideal" level—tending in particular, as we have seen in the Gough Map's English castles and roads in Wales, to be "about the acquisition of territory."[64]

As if to underline this point, the Gough Map has a handful of features (some textual, some pictorial) whose natures are not factual but expressly symbolic themselves. These features do not merely fall in line with the basic argument so far presented—that this is a map whose cumulative effect is to collapse the distinction between king's body and Britain's territory in support of a Westminster-centered fiscal monarchy of the whole island. In fact, Gough's symbolic material goes much further than does even its predisposedly acquisitive factual material, in invoking support for English overlordship of all Britain.

KINGSHIP, PART FOUR: BRUTUS LANDED HERE

The Gough Map's operation on an ideal or paradigmatic level (in addition to a factual geographic one) may be seen most simply in its handful of references to a legendary Arthurian topography. "Something a little unusual" in the context of Gough as an aid for travelers, for example, is the map's "prominent marking" of *The Wathelyn*, a lake (one of just four on the map) east of Carlisle in Cumberland. In two late Arthurian romances, both produced in England's northwest (*The Awyntyrs off Arthure at the Tarne Wathelyn*, c.1400; *Sir Gawain and the Carl of Carlisle*, c.1425), the Tarne Wathelyn figures as a site of otherworldly contact, a magical lake of transformative powers. Outside romance, however, no wayfarer has yet encountered it.[65] It comes as no surprise that Glastonbury, where at Easter in 1278 King Edward and Queen Eleanor witnessed the opening of Arthur's Tomb, should appear on the map, given the wealth and importance of the abbey there. But two other more expressly legendary Arthurian sites also show up on Gough: Pendragon, the ruined castle of Arthur's father Uther Pendragon in Westmorland; and Tintagel, site of Arthur's magically assisted conception, represented on the map by a castle on an island off the Cornish coast near Boscastle.[66]

A far more explicit and provocative reference by the Gough Map to legendary British geography comes in a brief inscription off the coast of Devon: *hic Brutus applicuit cum Troianis* ("Here landed Brutus with the Trojans"). This invocation of prehistory establishes the imperial patrimony and destiny of Britain's kings. As Geoffrey of Monmouth's first book relates, after being exiled from Italy, Aeneas's great-grandson Brutus encounters a group of enslaved former Trojans in Greece, then leads them to a promised island ("beyond the setting of the sun, past the realms of Gaul") where they establish *altera troia* ("a second Troy"). "A race of kings will be born there from your stock," Brutus is told in a dream, "and the round circle of the whole earth will be subject to them." No drawing of Trojan ship or eponymous hero adorns the coast, and no lengthy text (such as a *mappamundi* or Matthew Paris map or *Brut* chronicle might include) celebrates his coming ashore at Totnes.[67] Instead, in its unsensational presentation alongside what is otherwise factual geographic data, this gloss elevates Geoffrey of Monmouth's account of Trojan-Roman-British *translatio imperii* from the level of sheer mytho-propaganda to that of sober topographic description.

Geoffrey's wildly popular *Historia* (over 200 manuscripts survive) had been current in England for a century and a half, angled from the beginning toward a royal readership.[68] But there is a better explanation than just this text's general currency and relevance to English kings to account for the Gough Map's notation of Brutus's landing. Edward I's Welsh wars were only a first expression of his evolving ambition to overlordship of all of Britain. Scotland, a much different case in being a kingdom unto itself, was the larger prize and far more difficult to secure than Wales: Edward died on campaign while pressing his claim in 1307, bequeathing to son and grandson an ambition they would never see realized.

Long the norm in Scotland's relationship to the Anglo-Norman monarchy, already by the reigns of Edward I of England (1272–1307) and Alexander III of Scotland (1249–1286), had been a degree of "political deference," a quality underlined by England's overwhelming advantages (real and perceived) in social, cultural, economic, and military affairs. It is important to be clear that medieval processes of domination and even programs of military conquest were only rarely the product of "political masterminding," or "part of some grand strategy"—though it has been suggested that Edward I may provide something of an exception to this. Generally speaking "kings seized their opportunities as they arose," and true to this observation Edward did not move against Scotland either legally or militarily until provided with a good opportunity to press his rights (indeed, depending on the precise terms of his coronation oath, he had arguably a responsibility to do so).[69]

This opportunity came in the succession crisis which resulted from the death in late September 1290 of Margaret, the *domicella Norwagie* or "Maid of Norway" who had been accepted as Queen of Scotland following the death of her grandfather Alexander III in 1286.[70] In July 1290 plans were laid for a union of the kingdoms of Scotland and England through betrothal of seven-year-old Margaret to six-year-old Edward of Caernarvon, with dispensation from Rome already secured and with the endorsement both of the Scottish guardians and of Margaret's father, King Erik of Norway. From this point through until his death in 1307, Edward I found himself deeply involved with the affairs of Scotland; over time he became "more and more absorbed," even "obsessed with the desire to assert and to maintain his alleged rights." If it may be said that in arranging his son's betrothal to Margaret Edward was in effect already pursuing overlordship of Scotland diplomatically, in the affair's next phase (i.e., following Margaret's death) he moved legally, at a series of hearings in 1291 which established John Balliol as King of Scotland but under English overlordship. Later Edward took to pressing the matter militarily, in a series of campaigns against first Balliol (1296), then William Wallace (1297–1305), and ultimately Robert Bruce the Younger (1306 and ff.). Yet it would be misleading to suggest that from the beginning Edward's sights were "set on securing Scotland, as the dearest ambition of his closing years."[71] In 1290 Edward I was preeminent among the nobility of western Europe; the only kingdom worth his conquering, as E. L. G. Stones and G. G. Simpson have said, was Jerusalem, and to this end he had in 1287 duly "taken the cross" for a second time. Following 1287 English records are littered with correspondence between king and pope on the subject of Edward's planned departure, especially its date (pushed back twice already by 1290) and, even more delicate, its financing. But the Maid of Norway's unforeseen death reconfigured Edward's course significantly, so that his crusade came to be deferred repeatedly (in the end, permanently) as he dealt with the more pressing matters nearer home of Scotland, as well as another Welsh rising in 1295.[72]

Since the eighteenth century if not before, the legal dispute over Scotland's throne has been known as "The Great Cause." The defining period in the affair consisted of a series of meetings held at Norham and Berwick beginning in May 1291. Here a roster of claimants to the throne led by John Balliol and Robert Bruce the Competitor presented their cases, but only after first agreeing to Edward I's adjudication and to the principle of the English Crown's right (as *superior dominus*) to overlordship of Scotland, *la sovereine segnurie*.[73] "Thanks to the evident desire of King Edward to place the proceedings on record," the Great Cause has left behind a great mass of documentation, including two full notarial records of the 1291 proceedings, as well as a handful of partial accounts and miscellaneous related material. As soon as all competitors had agreed to his condition of suzerainty, for example, Edward dispatched a letter to more than two-dozen English religious houses requiring that they record this submission in their chronicles.[74]

More famously, in the spring of 1291 Edward I issued the first of two celebrated appeals to English monasteries for historical material he might use to establish his right to overlordship in the upcoming Great Cause hearings. A privy seal writ dated the eighth of March requests the Abbot of Evesham (the royal household was nearby) to "examine his chronicles" for anything "touching in any way our realm and the rule of Scotland" (*regnum nostrum et regimen Scocie qualitercumque contingencia*). It remains unclear how many monasteries were written to in 1291, but responses survive from twenty houses, and others (especially the greater houses) may have sent representatives rather than letters. Evidence suggests that some monks, bearing books of history with them, were present at Norham for the Great Cause's opening hearings in May, providing "considerable value as propaganda, whatever the historical value of the evidence itself."[75]

The notarial rolls make it clear that royal clerks prepared a historical survey as a brief for the occasion, to be read out in order to establish Edward's claim to jurisdiction. In doing so, they appear to have worked primarily from their own archival researches and to have looked only thereafter to incorporate material obtained from the monasteries. Time was short, the request belated, and the monastic returns actually of little use: many bear comments such as *nil novum invenitur* or *nichil continet ad propsitum* ("nothing new is found"; "nothing to the purpose"). Several returns to the royal inquiry included reference to the legendary history of Geoffrey of Monmouth, but in 1291 Edward's clerks "fought shy" of this material and produced instead a compilation which began only in 901.[76] This apparent skepticism toward the Brut legend (when there was much to profit by in it) is somewhat surprising, considering not just Geoffrey's astounding popularity but more especially "the credence he was given by reputable and scholarly writers."[77]

A decade later Edward issued his second appeal to history. By this time all freely given recognition of an English right to overlordship of Scotland had long since ceased, and John Balliol (installed 1291, defeated and deposed 1296, brought captive to England and eventually exiled to Normandy) had been superseded by William Wallace, who in 1297 rose to power and ruled in Balliol's name. After Wallace won a memorable victory at Stirling Bridge, English war-subsidies and enlistment increased sharply and the matter of Scotland "dominated the political scene" for the rest of Edward's reign. Edward I's second request for historiographical help from his religious houses thus came in the midst of a full-scale imperial effort; specifically, it was formulated in response to a 1299 bull of Pope Boniface VIII "demanding that the king abandon the war in Scotland" (*quod rex desistat a guerra Scocie*) and claiming that kingdom as a papal fief. Boniface's letter to Edward contends that "from ancient times the realm of Scotland belonged rightfully, and is known still to belong, to the Roman church," and "was not, and is not, feudally subject to your predecessors, the kings of the realm of England, nor to you." However, none of the pope's historical material (apparently furnished by the Scots) reaches earlier than the reign of Alexander III (1249–1286).[78]

Edward's second inquiry to English religious houses was "better organized and less hurried" than the first. No written returns exist, but the distribution-list to the king's letter (this time less vague in its terms) includes upwards of 45 abbeys, priories, and secular cathedral chapters; royal archives were also searched. Despite the trouble taken, little new material seems to have been secured. The primary difference between the argument as presented at the Great Cause in 1291 and that forwarded in Edward's 7 May 1301 reply to Boniface is that the compiler of the latter ("a man of considerable skill") abridges the historical survey significantly while yet expanding its scope. Thus on one end it continues through Balliol's 1296 rebellion and deposition rather than concluding in the mid-century (as the Great Cause version did), hence counteracting the pope's arguments concerning the reign of Alexander III. Moreover on the other end, the historical argument made in 1301 begins not merely in the tenth century, as had the 1291 version, but with a "mythological prologue" treating Brutus and his eldest son Locrinus, then Dunvallo Molmutius and his eldest son Belinus, and eventually that *princeps famosissimus* Arthur, "prince most renowned," who following the precedent of each of these forbears "subjected to himself a rebellious Scotland," after which "in succession all the kings of Scotland have been subject to all the kings of the Britons."[79]

The source for this, of course, is Geoffrey of Monmouth. His bestselling mythography had been spurned a decade previous but by the end of the century the *Historia*'s account of the "passage of dominion" from Priam's Troy down to Norman England had come to be recognized as providing Edward I with his most effective pretext for the "reconquest" of Scotland. In their definitive account of the Great Cause, Stones and Simpson note the "somewhat skeptical attitude" of the Scottish respondents to these claims; nonetheless, the Scots sought to better the English at their own game by way of an alternative legendary history, in which a daughter of Pharaoh named Scota acquires an independent Scottish realm through conquest of one of Brutus's sons (all of whom, they argue incidentally, Brutus had made "peers, so that none of them were subject to another"). In any case, the great stock which England's royal house placed in Geoffrey's version of the story may be inferred through another event from 1301: in a ceremony at Caernarvon marking his accession as Prince of Wales, the future Edward II was presented by his father with a copy of the *Historia Regum Britannia*, perhaps meant to serve as a kind of *Fürstenspiegel* ("mirror of princes") or handbook of his hereditary territorial rights.[80]

The Gough Map's modest legend off the coast of Devon thus takes us out very decidedly from the realm of mere wayfaring utility. Five words, not native to the original map (i.e., the posited prototype of c.1280) but those of some early glossator or intermediary compiler (c.1300–02?), have reoriented once more the basic thrust or tenor of the document, further transforming our characterization of its ideal, prototypical user from one concerned simply with effective travel, into one concerned not only with the stewardship of the royal fisc or public domain, but with the effective alignment of all this sphere's rights and properties, all the realm's resources and components accessible on behalf of the common good, to the specific ends of Welsh subjugation and Scottish conquest. These achievements would provide a benefit pertaining to all English society, but could only be accomplished with the approbation of all, and through the participation (monetary, military, and otherwise) stemming from this. All Droitwich's salt and all Grimsby's cod, not to mention Wales's archers and Southhampton's navy, would be needed to help put Brutus's island-realm back together again.

The Gough Map is in fact not alone among thirteenth-century maps in being implicated in (or at least connected with) Edward I and his administrators' Great Cause or attempted subjugation of Scotland. After the two full notarial rolls, the next most important documentation of the 1291 hearings and their aftermath is a source known as the *Annales Regni Scotiae*. Sometime before 1325 (and "probably somewhat earlier") this abbreviated and partly secondhand account of the Great Cause proceedings was copied into B.L. MS Cotton Claudius D.VI—a St. Albans manuscript that included among its preliminary material one of Matthew Paris's four Maps of Britain (in addition, incidentally, to the sketch with reference to the king's *mappamundi* so important to our discussion in chapter one). Matthew's Claudius version of his Map of Britain is generally regarded as his best overall, offering (perhaps not coincidentally) his most detailed topographical representation of Scotland, together with his fullest textual attention to it, in terms toponymic and otherwise (see fig 2.7, above). Judging by this manuscript's pre-drawn columns and colored borders, Cotton Claudius D.VI appears to have been prepared especially for Matthew's use, yet was left half-unfinished by him, to be returned to some fifty years later and utilized after the manner of the *Liber Additamentorum*, Paris's own *Chronica Majora* appendix. That is to say, in the space following Paris's work this manuscript is filled up with miscellaneous items such as copies of official documents and scraps of chronicles, as well as related material including the c.1300–1305 political poem *De Bello Scotico*.[81]

The original unity of the manuscript is not in question. It is therefore clear that from c.1250–55 onward its Map of Britain stood between a series of drawings (ff.5v-9r) of seated English monarchs known as the "Genealogy of Kings" (fig. 1.27, above) and Matthew's *Abbreviatio Chronicorum* (10v-98v). This means that when further material (most of which treats the theme of Edward I and his relationship to Scotland) was added by St. Albans scribes a generation later, beginning probably c.1302–08, these autograph headline items of abbreviated chronicle, line of kings, and map of the island had long been securely in place.[82] Doubtless, they served for subsequent compilation as markers of the volume's thematic character.

The choice to copy a collection of Great Cause notarial material and related chronicle fragments into such a manuscript, one prefaced by the British genealogical, historiographic, and cartographic authority of Matthew Paris himself, may well owe something to the precedent offered by an existing version of the Gough Map. Or alternately, an encounter with the Claudius manuscript may have persuaded some early glossator or copyist to add to a Gough template in the royal archives the relevant note of Brutus's landing—relevant, that is, to a medieval audience, if initially perplexing to a modern one.

In addition to a shared Great Cause context, there are some provocative parallels between the Gough Map and Matthew's Claudius Map of Britain in terms of their representation of Scotland. For example, the island of Orkney's marked displacement to the south on Gough has a direct antecedent in Claudius—in fact the Orkneys appear off the Firth of Forth on this version alone among Matthew's four. The Claudius Map of Britain also appears to provide precedent for an unusual extra-topographic feature on the Gough Map, a legend in Sutherland noting that *hic habundunt lupi* ("here dwell wolves"), written below the drawing of a wolf set among a crowd of scalloped hills (fig. 3.9). As Parsons has noted, this is "no doubt the *regio montuosa et nemorosa*" ("region full of mountains and woods") which Matthew Paris's Claudius and Corpus Christi Maps of Britain describe in multi-line inscriptions beneath the counties of Caithness and Sutherland (see fig. 2.7, above).[83]

To have procured for copying at St. Albans the official Great Cause notarial material comprising much of the *Annales*, in the first place, will have required a monastic-chronicler with a source well placed in the royal administration. But the choice to copy such records into a manuscript prefaced by a genealogical line of English kings stretching back through Arthur to Brutus, together with a map of the whole island, argues the full blending in this period of a cartographical consciousness with not only a historical and administrative one, but one indeed of explicitly imperial contours. In short, partly by way of some cartographic detail but even more so through the other items compiled into its manuscript, Matthew Paris's Claudius version of his Map of Britain provides a model for the connection of an English regnal map to a project assigned increasing importance by royal administrators as the reign of Edward I wore on: that of securing and maintaining authority over Scotland and Wales. Once again, a medieval map proves inseparable from an essentially imperial context—from the discourses of conquest that frame it both conceptually and materially.

A BATTLE IN THE SEA, OR, SHIPS AND SOVEREIGNTY

Three final features of the Gough Map, each of them symbolic or pictorial in a manner all but alien to the topographical material around them, require analysis in the context of this document as an argument for Edward I's claim to overlordship of Scotland and for England's ancient rights to a monarchy of the whole island. To date, no single unified interpretation of all three of these features, nor any combination of discrete readings of them, has managed to win critical acceptance. We have seen that via a series of details added c.1283–95 to what seems to have been a c.1280 prototype—additions made in what is primarily a "factual geographic" mode: i.e., new castles, new roads—the Gough Map registers, and may have helped enact, the Edwardian conquest of Wales. The readings presented below dovetail similarly—in support of the proposition that sometime in the first half of the fourteenth century Gough was again provided with new material, in this case not measured roads or other internal details but instead marginal images operating in what may be described as an ideal mode, and serving this time in order to figure (or once again help enact, by winning approval for) English conquest of Scotland.

The first of Gough's special pictorial features is an elaborate line drawing of a wrecked ship, in size about as long as Wales or Scotland is wide (fig. 3.10, left). This ship faces south (left to right) in the North Sea halfway between *Norway* and the oversized *Insula de Orkeney*, but seems to have wavered somewhat from its course, to run disastrously aground on a rocky shoal (which had apparently been unforeseen, given its shading with the same watery-green as the rest of the Britain's encircling ocean). Several details in the construction of Gough's shattered vessel—e.g., its single mast, shown broken on the map—establish that this is a documentary drawing based on an English warship from about the second quarter of the fourteenth century, a point which underlines its likely absence from the map as originally compiled. Two rectangular rafts or biers are set before this ship, one of them empty, still caught on the rocks. The other, further out to sea from the wreck but closer to Orkney, holds "a stretched-out figure with long hair and clothed in a loose-fitting garment drawn in at the waist."[84]

In the space directly to the right of this drawing is Gough's second unusual pictorial feature, another boat, similar to the first in its size, shape, and orientation on the sea, in which a single oversized figure seems to bend toward the water in order to push out, or pull in, an object approximately the size and shape of one of the other drawing's biers (fig. 3.10, middle).[85] What most characterizes this second image is that it remains unfinished—indeed, it seems only just begun—unless possibly the figure was so heavily encrusted that after many years of hanging it has fallen away from the vellum, or has been pried off for the richness of its illumination, leaving behind not even a penned outline (such as limns every island, coastline, and topographical icon on the map) but rather just an empty, indistinctly delineated white space in the midst of green sea.

Gough's final unexplained illustration is a detailed rendering of three large fish or sea-creatures, locked in apparent combat, one against two, in the open space directly above an exaggerated wedge marking the Firth of Forth at the border of Scotland and England (fig. 3.10, right). Wholly unlike the second image and much beyond the extent even of the first, this third drawing in Gough's North Sea is artistically ambitious, complete in both outline and internal detail plus given shadings after the naturalistic manner (if not quite the mastery) of Matthew Paris's celebrated *Liber Additamentorum* elephant (a gift to Henry III from Louis IX of France in 1255; fig. 3.11). W. B. Sanders has described the scene as "a gigantic representation of the Whale engaged in mortal combat with its natural persecutors, the Thrasher and the Swordfish."[86] The largest creature, Sanders's Whale, hovers above England, its long body angled downward and to the left (i.e., facing northwest), in other words oriented just as is the Scottish March below it (fig. 3.12). The next largest fish, Sanders's Thrasher, faces this one from the north and has a rocky back, with four scales jutting up sharply to match the similarly four-scalloped hills of the Scottish highlands (fig. 3.13). The smallest fish, Sanders's Swordfish, has much the same shading as the largest (the Whale), but is placed almost parallel to the Thrasher, set against their common aggressor from the west and north.

The prevailing although generic speculation is that Gough's sea-creatures may be "symbols of some incident in history." Well worth noting in this vein is the entry in Matthew Paris's *Chronica Majora* for the year 1240, in which "an unprecedented battle among the fish, whales, and marine monsters" off the coast of England is implied to have dire implications for human events, or else to have come about as a consequence of them. In fact 1240's sea battle, as Suzanne Lewis has observed, serves as "the culmination of a whole series of natural and human disasters," in a year in which "the course of events" (so Matthew's entry begins) "was unfavorable to the kingdom of England." Beside this text, in an "elegant but terrible image" more intense than but not totally dissimilar from that on the Gough Map, three frenetic multicolored *belue marine* ("sea-monsters") are locked in bloody combat upon a light-green sea, while a fourth has all but disappeared down a throat (fig. 3.14). Unlike Gough with its variety of species and discrete battle-stations, the *Chronica Majora*'s sea-monsters are all of a single type, and desperately interlaced. This same folio includes a report on the Tartars, whose ravages on the frontiers of Europe were seen by many as an unleashing of Gog and Magog, a sign of the coming of the Antichrist. A few folios later comes a description and panoramic illustration of 1241's "momentous naval battle" between the imperially aligned Pisans and the papally aligned Genoese, in which the former won a decisive victory on behalf of Frederick II—who in 1241 was himself rumored to be in league with the Tartars against Christendom. In the end, "the reader is left to judge" (Lewis has said) between such candidates as these, in determining the meaning of the *Terribile prognosticum de piscibus in mari pugnantibus* ("Dreadful sign of fish fighting in the sea"), as the section's oblique rubric runs. But whatever its precise historical resonance, such a *bello in mari* ("fight in the sea") as this clearly has major import for contemporary human affairs, the clash of *beluae marinae* seeming to resonate in particular with the clash of armies.[87]

Given all that has come before, the Gough Map's three sea-creatures' relative sizes, spatial disposition, and antagonistic orientation (one larger from the south against two smaller from north and west) will surely have suggested already what "incident in history," what clash of armies this scene might represent. And in fact, the alliance of Thrasher and Swordfish against the bullying Whale looks to be a particularly appropriate narrative detail. On one hand, it is true that Edward I employed numerous Welsh troops in his campaigns against Scotland, including some 10,500 archers at the 1298 Battle of Falkirk alone; and of course by no means did the Scots leap to the defense of Snowdonia in the years of Welsh subjugation. However, in 1295 rumors began to circulate (in association with the treason of Thomas de Turberville) to the effect that Wales and Scotland would rise together to throw off English tyranny—and indeed contemporary documentary evidence does suggest the stirrings of a Welsh/Scottish coalition.[88]

Neither whale nor thrasher carries any very particular symbolic resonance in medieval art; however, the wide semantic field tapped by the Latin *gladius* ("swordfish") offers much in the way of relevant material. Even on a concrete level, the term denotes not just a species of fish—more often it means simply "sword," and occasionally "fleur-de-lit" (an ellipsis of *gladiolus*). D. R. Howlett's *Dictionary of Medieval Latin from British Sources* (1989) provides a string of intriguing conceptual definitions to add to these material ones, from "warfare" and "country of origin" (e.g., 1288's *cum gladio Scotio*) to "symbol of knighthood" and "symbol of authority, lordship" (especially "royal").[89] The term *gladius*'s strong political and military associations, in short, make the icon of the swordfish an appropriate emblem for a contestant in a battle over the extent of

Edward I's royal lordship in Britain.

In 1895 Sanders outlined the suggestion that Gough's shipwreck scene represents "the fatal end of the calamitous voyage" undertaken in Autumn 1290 by the young *domicella Norwagie*, Margaret, Queen of Scotland. Parsons looked into Sanders's "interesting theory" in his seminal introduction to the R. G. S. facsimile version of Gough sixty years later, but judged it untenable on the grounds that although "there is some evidence of the Maid's arrival in Orkney and death there, there is none of a shipwreck." In place of the Maid of Norway's voyage, Parsons suggests that what may have "inspired" Gough's drawing is an event from the Old Icelandic *Orkneyinga Saga* (c.1200), in which Earl Rognvald of Orkney is shipwrecked c.1150 at Gulberwick in the Shetland Islands during a stormy crossing from Norway. However, numerous factors, among them the Shetlands' absence from Gough (even though they lie within the map's compass) and the great unlikelihood (to say the least) of any special attachment on the part of an English royal clerk or other user of the map to a minor scene in an obscure saga in a language all but unknown in English libraries c.1280–1360, make Parsons's proposition virtually impossible to sustain, less tenable by far than Sanders's.[90]

Rather than posit an improbable link to an irrelevant text in order to explain Gough's North Sea shipwreck, it seems more advisable to depend on the considerable role rumor seems to have played in the limited understanding which contemporaries had of the circumstances surrounding the Maid of Norway's death in 1290. In October Scottish envoys prepared to set out for Orkney to receive Queen Margaret, who once in their custody was to be brought for inauguration to Scone, near where the realm's nobles and prelates had assembled at Perth, afterwards to be married to the future Edward II. Thus an English monarchy over the whole of Britain seemed on the verge of materializing; once achieved, Edward I would turn full attention to his departure on crusade, in preparation for which project English laity and clergy had been taxed heavily since 1287.[91] Beginning at the point of the Scottish envoys' intended departure for Orkney, however, considerable confusion ensues.[92] Far from settled questions throughout Autumn 1290 were not only the manner and the timing of the Maid of Norway's demise, but even the very question of it—was she convalescent in Orkney, after shipwreck or sickness, or already dead?

Rumors of the Maid of Norway's death first came south in a letter to Edward (dated 7 October) from William Fraser, the Bishop of St. Andrews, who had been about to depart for Orkney when report of Margaret's death came, followed by others suggesting that she had instead *convaluit de infirmitate sua* ("recovered from her illness" or "regained strength after her weakness"). Those with claims to an apparently vacant throne have begun, the bishop reports, to arrive at Perth (e.g., Robert Bruce *cum magna potencia*, "with a great following") such that war seems imminent, unless it might be averted *per industriam et ministerium vestrum* ("by your active intervention"). As one of the realm's four guardians, the bishop asks Edward, if rumors of Margaret's death be true, to come to the border *ad consolatium populi Scoticani, et ut effusioni sanguinis parcatur* ("in order to maintain the morale of the people of Scotland, and to prevent the spilling of blood"). Bishop Fraser's letter has been described as "hasty" and "uncertain" but nothing more definitive appears to have come out of Scotland for some time. Margaret was in fact dead by the 26th of September. However, not just sources immediately contemporary but also those treating the affair in years afterwards bear the mark, in their frequently misinformed facts, of this crucial early period of confusion.[93]

It remains unknown and may be immaterial precisely how Queen Margaret's life and sea journey ended. For the Gough Map's illustration of a dead maiden between Norway and Orkney tells the story of the Maid of Norway's "calamitous voyage" on a level more important than the strictly factual. The young queen's broken vessel and drifting bier, failed in the charge of transporting her unto sovereignty, may be read as part fact, part rumor, part metaphor—the in-any-case undeniable wreck, in a 1290 disaster at Orkney, of Scotland's fragile and wayward ship of state.

The rhetorical figure of the ship of state had been current in a general sense since classical times, having been invoked by Aeschylus, Plato, Sophocles, Theogenes, Horace, and Demosthenes, among others.[94] But what is more firmly established by the contexts of its medieval usage is the appropriateness of this image to precisely such a historical subject as the Scottish succession-crisis initiated by the Maid of Norway's death at sea. Two late-medieval English poems, *The Death of Edward III* (c.1377) and *The Ship of State* (c.1458), employ this metaphor as their central structural conceit—each delineating the figure of an endangered ship of state in order to comment on the turmoil involved in royal succession, with respect in particular to the anxiety-inducing prospect of an underage ruler. As the later poem laments, "Formerly we had a noble English ship, strong against storms, a defense of the land," one which "had a sure mast and a strong large sail"—but "all wears and washes away." Just as does the *Ship of State* (which "in form, style and even phraseology this poem closely resembles"), *The Death of Edward III* proceeds by setting out the various qualities and figures comprising the community of the realm and likening them point by point "to the several parts of a warship," which is to say King Edward III, an emblem for the kingdom.[95] Judging from this evidence, it would seem that the allegory of the ship of state in medieval Britain was connected with the issue of uncertain royal succession, especially that involving a minor. This makes it an apt image for the affairs of c.1290–1307.

Contemporary visual evidence likewise indicates that an identification between matters of state and the Gough Map's North Sea ship-illustrations may be extremely appropriate. In 1344 England began issuing a new gold coin, known as the Noble. This name, adopted immediately (the Noble is "referred to as such in mint records"), is apparently "drawn from the noble person portrayed on the coin's obverse": Edward III. However, where previous English coins (such as the Florin, issued January-August 1344) bear on their obverse a representation of king on throne (as on Henry III and Edward I's seals; see figs. 1.15–1.16, above), the Noble departs from convention to depict, instead, King Edward seated in a ship (fig. 3.15). This is the first appearance of a ship on any medieval coin.[96]

The king-bearing ship on the 1344 gold Noble is "a stylized design, rather more realistic than usual," which took as its model the fourteenth-century cog—that is, the "normal one-masted square-rigged vessel used for both trade and war" whose appearance we have witnessed once already, in the Gough Map's wreck off Orkney. The Noble's deployment of this ship-motif usually has been read as a commemoration of Edward III's important naval victory over the French at Sluys in 1340—*Sklus* being one of only five towns marked by the Gough Map on the continent, incidentally. But probably this coin's design serves a number of purposes, also marking, for example, England's general rise in these years in terms of its overall naval power. As recalled in the sixteenth century, this period's unusually expressive coins link "King, ship, sword and power of the sea," underlining the point that during the reign of Edward III, "in special/ the sea was kept, and thereat he was lord./ Thus he made he nobles coined as record."[97]

The Noble's curious enthronement of King Edward on a ship may also be read as claim to an extension of the bounds of English sovereignty. In a 1332 dispute with Genoese pirates over the king's right to seizure of shipwrecked goods, the legal advocates of Edward III employed precisely the arguments which in c.1260 Henry Bracton had formulated, regarding the crown's possession of "certain inalienable rights," but extended them to include the new assertion of an English "royal and imperial sovereignty of the sea." Just as the King of England had fiscal claim to an elaborate roster of sovereign rights within the territorial bounds of his kingdom, so too, comparably, did he enjoy "sovereign lordship of the sea of England, with all its islands and belongings," the realm's surrounding waters being considered to constitute its "sea-frontier." Central to such lordship was the king's "right of wreck" along the coast.[98]

All signs indicate that the ships in Gough's North Sea must have been drawn, at least in their present form, sometime around the middle of the fourteenth century. And in line with this, we have seen that marking the reign of Edward III and in particular the 1340s is a new concern for the relationship between English royal sovereignty and the sea. However, just as surely as did the reign of Edward III, the reign of Edward I had its anxieties about the status of England's coastlines and surrounding waters; and as in the 1340s, so in the 1290s did royal officials take steps to secure the sovereign maintenance of the realm's sea-frontiers.

In 1295 English fear of invasion was widespread, following a French raid on Dover. The prospect of a combined French-Norwegian attack, coupled with the increasing intransigency of Scotland and a new Welsh rebellion (this is precisely the moment at which rumors of a coordinated rising appear), meant that England's coasts were suddenly "key to Edward's military problem." Accordingly, in August 1295 Edward instituted a new coastal defense scheme, in which Keepers of the Seacoast were appointed, to watch the eastern shore against Norwegians and Scots, and the southern against the French.[99] Events such as 1332's legal clash between the Crown and Genoese pirates, 1341's naval victory at Sluys, and 1344's adoption of the Noble help link the image of a king sitting in a ship directly to such affairs as bureaucratic state-formation and English imperial ambition, concerns which we have likewise seen to have peculiar relevance for the Gough Map. It is not too much to say that here in 1295, on the eve of the first of Edward I's campaigns against the Scots, naval affairs had already become a extreme source of English anxiety, a matter for the king's clerks' close attention. In time the sea would become a noted source of strength, with much of the military success achieved against Scotland during the ensuing half-century appearing to have its basis in English naval supremacy.[100]

If King Edward III enthroned in a ship on the obverse of a gold Noble constitutes an assertion of English sovereignty of the sea, what does the Gough Map's contemporary and closely comparable image, its shadowy negative of an oversized figure in a ship alongside the wreck of a cog off Orkney, then signify? In the context of the Great Cause, this outline-remnant or abandoned beginnings of a drawing of a mariner hunched over an object half-in, half-out of the water, looks to have a compelling hypothetical explanation. If to its left, to the north, is Scotland's succession-wracked ship of state, with its sovereignty in the prone figure of the Maid of Norway set adrift in the desperate waters of a *guerra communi* ("civil war") stirred to tempest by those in contention for the throne (so the Bishop of St. Andrews), perhaps this illustration has appropriately been removed or left incomplete. For what it seems poised to mark is the English monarchy's timely rescue, in the figure of a bending King Edward, of the drifting bier of Scotland's royal house, its maiden cargo set until so recently to have been gathered in and hauled up (via dynastic marriage) onto the secure deck of England's ship of state, sure of mast, strong of sail, and steered by an old crusader "feared throughout Christendom." Thus *per industriam et ministerium*—"through an active intervention," and one justified by the Bishop of St. Andrews's initial invitation to adjudication; by the Scottish claimants opening acknowledgment of English overlordship; by the historical precedent provided in Brutus and Arthur; and perhaps metaphorically, by his "right of wreck" in the sovereign seas around his kingdom—has the King of England responded as *superior dominus* to the political crisis generated by the shipwreck of the Scottish monarchy's fortunes.[101] King Edward bends to offer a correction to Scotland's course, rudderless since the death in 1286 of his peer and friend Alexander III.

At the beginning of his short and disputed reign, John Balliol was exhorted "to rule justly," lest Edward who put him on his throne "should have cause to intervene."[102] In its North Sea scenes, I believe, the Gough Map narrates an argument justifying ongoing English intervention into Scotland—although the claim that English aggression constitutes a benevolent rescue of the Scottish ship of state seems over time to have been rescinded. For in practice, matters between the self-styled senior kingdom and its recalcitrant junior were nowhere as clear as this narrative of benevolent intervention would suggest. The Battles of Stirling Bridge (1297), Falkirk (1298), Bannockburn (1314), and the rest make it clear that Edward's intervention hardly worked *ad consolacionem populi Scoticani*, "to console the people of Scotland" as Bishop Fraser intended—instead it brought about the *magna stange hominem* or "great massacre of men" he sought to avert.

This disjunction need not surprise us. Like other imperial documents, the Gough Map of Britain represents not just the *fait accompli* of domination, but includes rather a considerable degree of administrative ambition. According to all signs a document suited to English royal clerks—in terms of both compilation and use—this map did not just help the Crown's servants get around, the better to procure salt and archers for the king's domestic British wars. Rather, the Gough Map seems to have enabled these agents of an emergent English imperial bureaucracy, by and large "propagandists for royal authority," to travel to new places in political and social theory and practice, to articulate new and more highly laicized conceptions of civil and political association than had been possible even a short time before. Gough worked to foster in short "the development of a society in which primary allegiance is given to lay governments" as opposed—in the worldview typically found constitutive of medieval *mappaemundi*, for example—to the universal and transcendental church.[103]

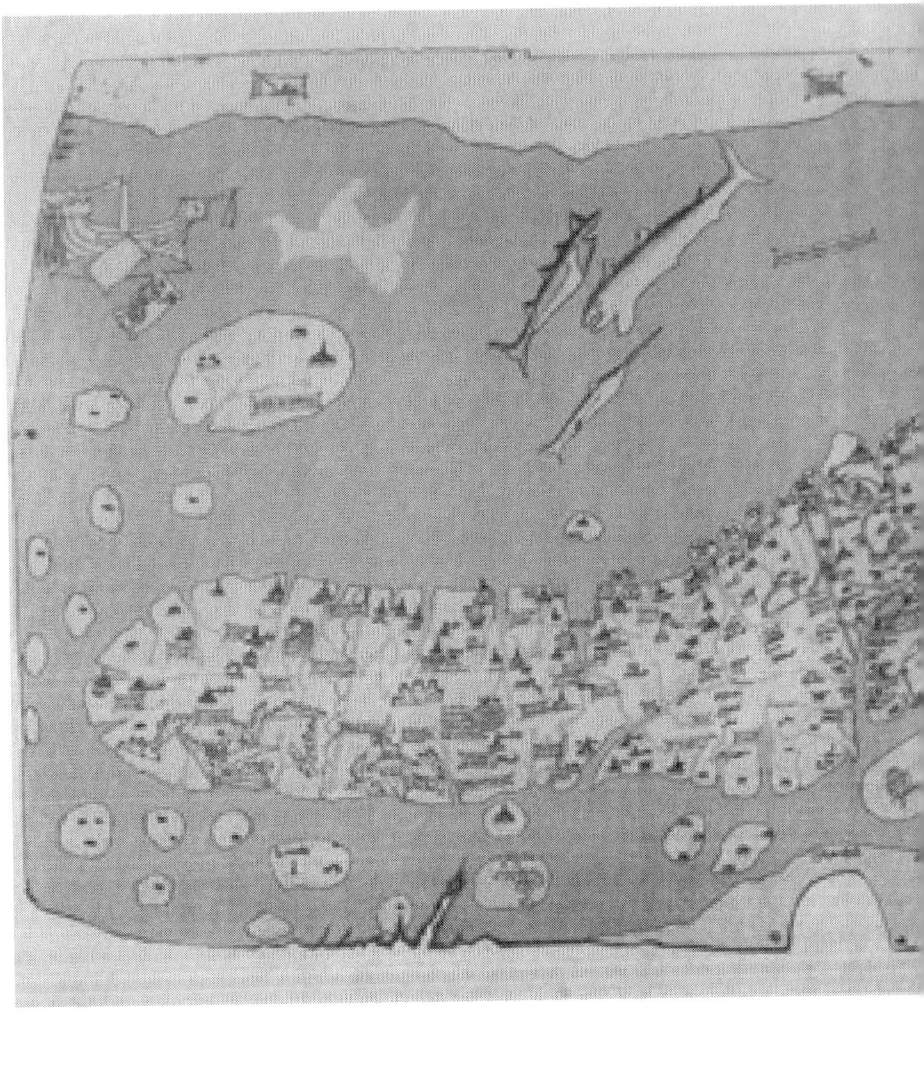

Figure 3.1. The Gough Map of Britain, c.1280–1360; Oxford, Bodleian Library Gough Gen.Top.16; RGS Facsimile Version, from Crone (1961); reproduced by permission of the Bodleian Library and the Royal Geographical Society.

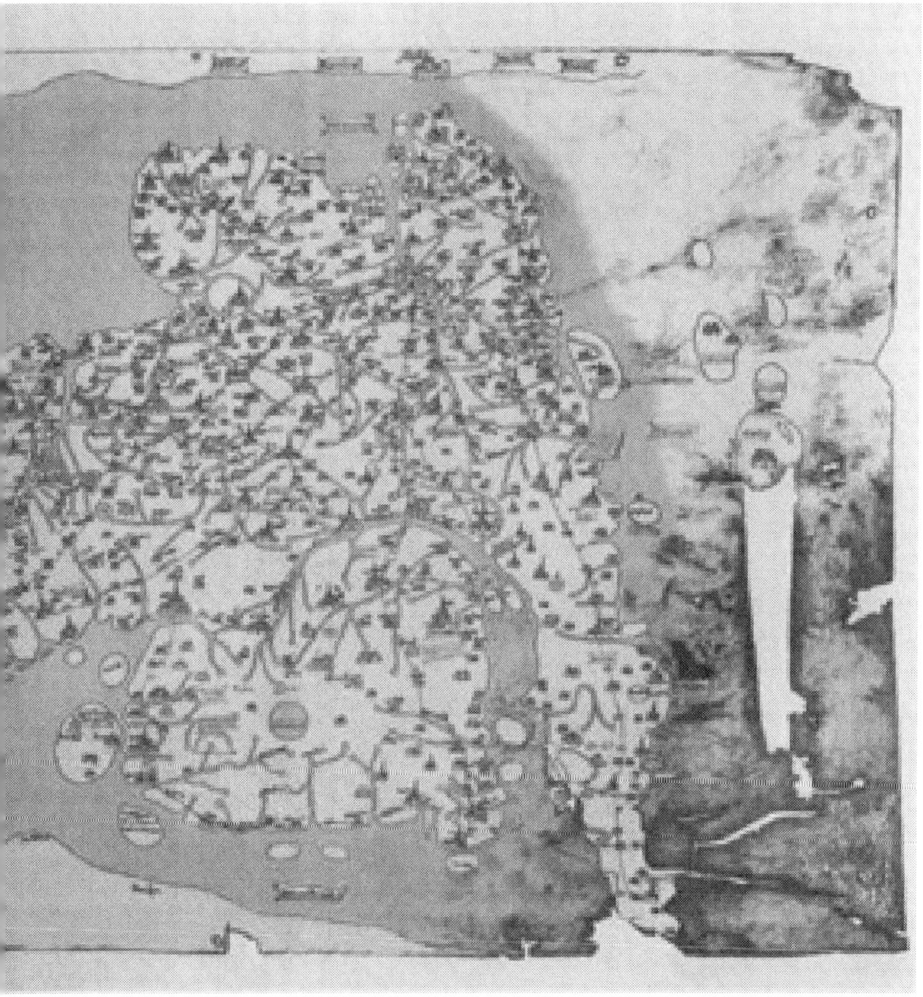

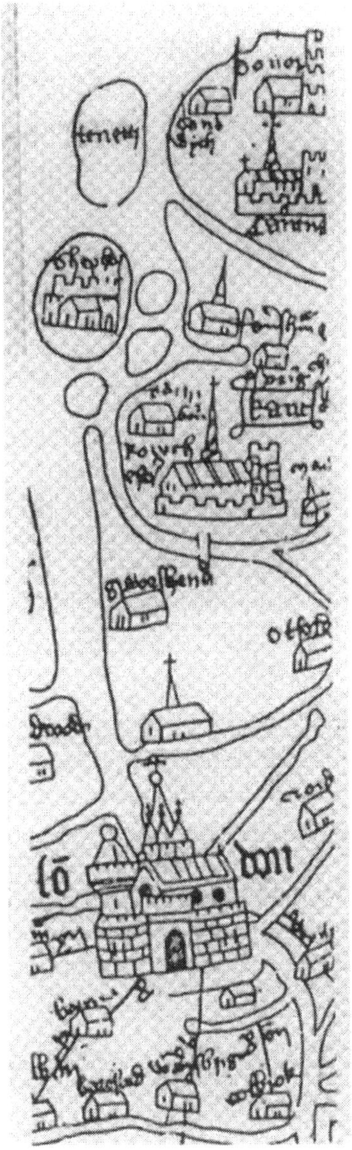

Figure 3.2. Thoroughfare from London to Dover, but no line for a road; detail from the Gough Map of Britain (facsimile version), c.1280–1360; note vertical alignment of towns and telltale bridge between Gravesend and Rochester; note also the various lines marking roads departing from London; reproduced by permission of the Bodleian Library and the Royal Geographical Society.

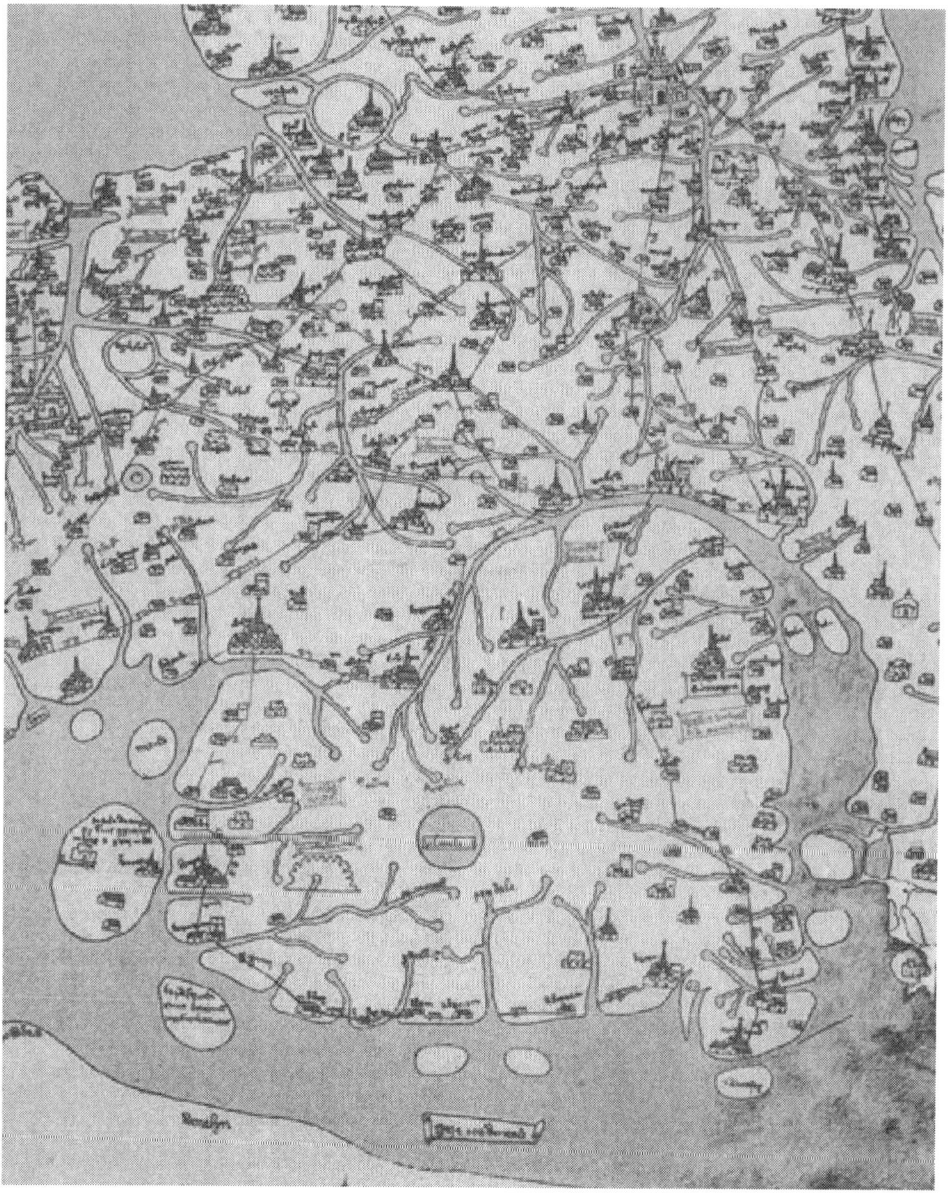

Figure 3.3. Wales, detail from the Gough Map of Britain, c.1280–1360; Oxford, Bodleian Library Gough Gen.Top.16, RGS facsimile version, from Crone (1961); reproduced by permission of the Bodleian Library and the Royal Geographical Society.

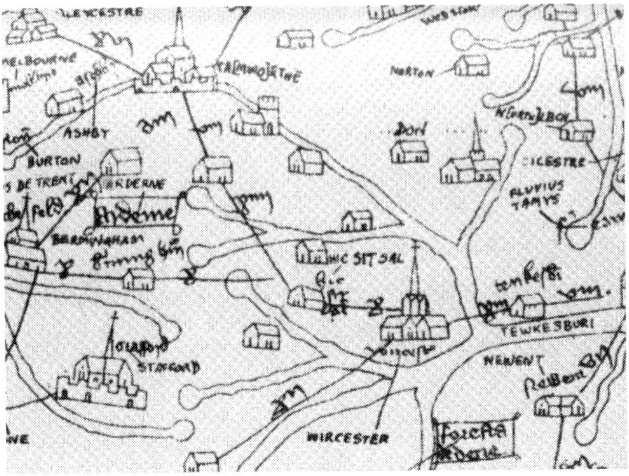

Figure 3.4. Droitwich, detail from the Gough Map of Britain, c.1280–1360; Oxford, Bodleian Library Gough Gen.Top.16; RGS facsimile version, from Crone (1961); note legend *hic fit sal* ("salt is produced here"); reproduced by permission of the Bodleian Library and the Royal Geographical Society.

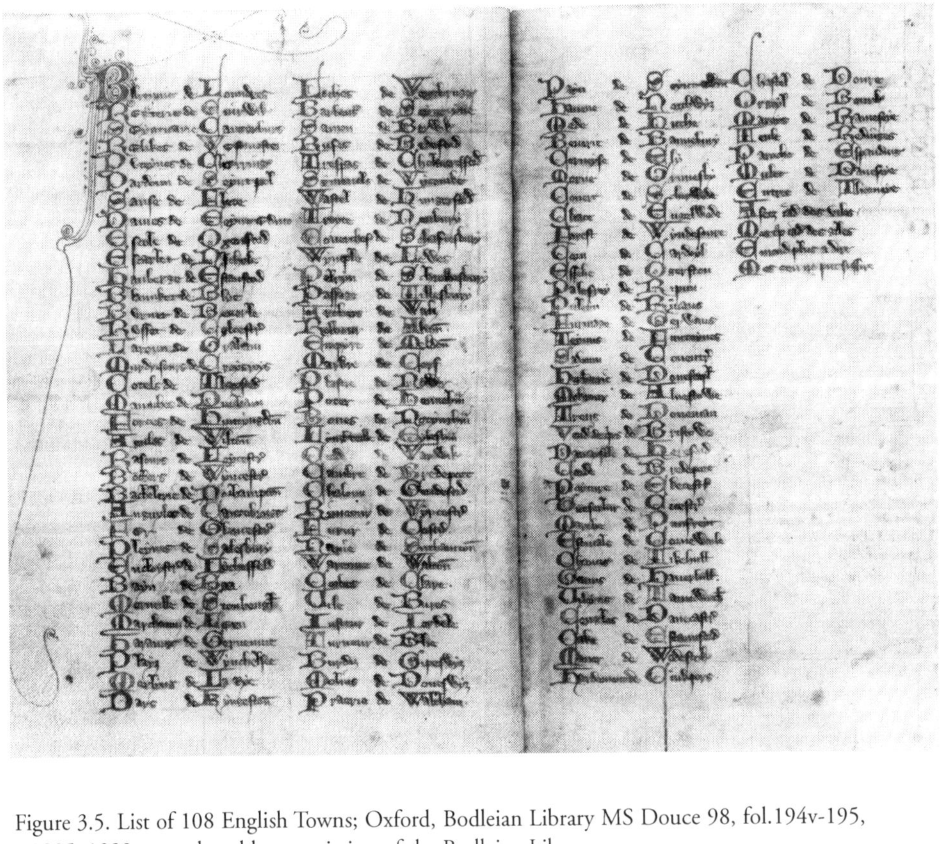

Figure 3.5. List of 108 English Towns; Oxford, Bodleian Library MS Douce 98, fol.194v-195, c.1295–1330; reproduced by permission of the Bodleian Library.

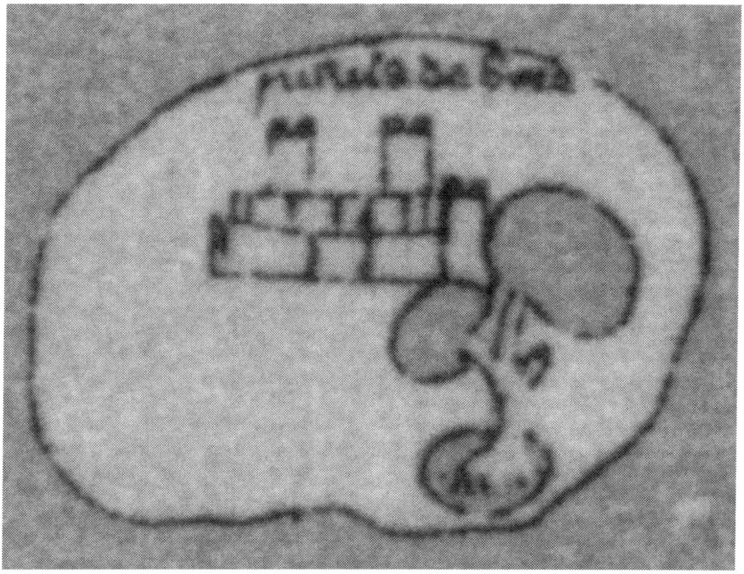

Figure 3.6. Icon of two intertwined trees on the Isle of Bute; detail from the Gough Map of Britain, c.1280–1360; Oxford, Bodleian Library Gough Gen.Top.16; RGS facsimile version, from Crone (1961); reproduced by permission of the Bodleian Library and the Royal Geographical Society.

Figure 3.7. Red Deer of Colgarth; detail from the Gough Map of Britain, c.1280–1360; note the legend *Colgarth hic maxima venacio* ("great hunting here") to the upper right of the deer, who looks set to flee deeper into the highlands; Oxford, Bodleian Library Gough Gen.Top.16; RGS facsimile version, from Crone (1961); reproduced by permission of the Bodleian Library and the Royal Geographical Society.

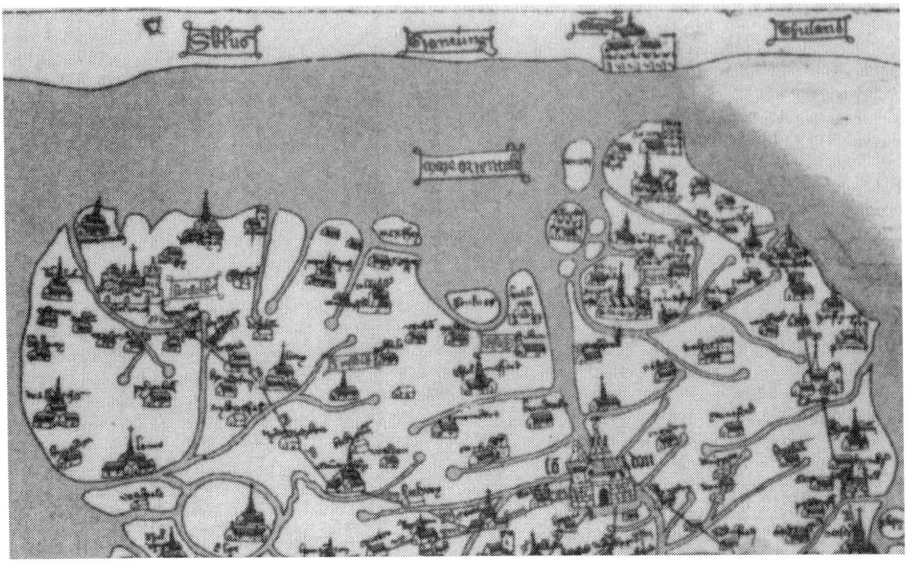

Figure 3.8. Southeast England and Calais; detail from the Gough Map of Britain, c.1280–1360; Oxford, Bodleian Library Gough Gen.Top.16; RGS facsimile version, from Crone (1961); note prominent walled town-icon for Calais but absence of icons for other towns on the continent, which are instead framed by cartouches such as mark districts in Britain; reproduced by permission of the Bodleian Library and the Royal Geographical Society.

Figure 3.9. Wolf of Sunderland; detail from the Gough Map of Britain, c.1280–1360; note the legend *hic habundunt lupi* ("here dwell wolves") beside the back leg; this detail appears just to the left (north) of Colgarth and its Red Deer; Oxford, Bodleian Library Gough Gen.Top.16; RGS facsimile version, from Crone (1961); reproduced by permission of the Bodleian Library and the Royal Geographical Society.

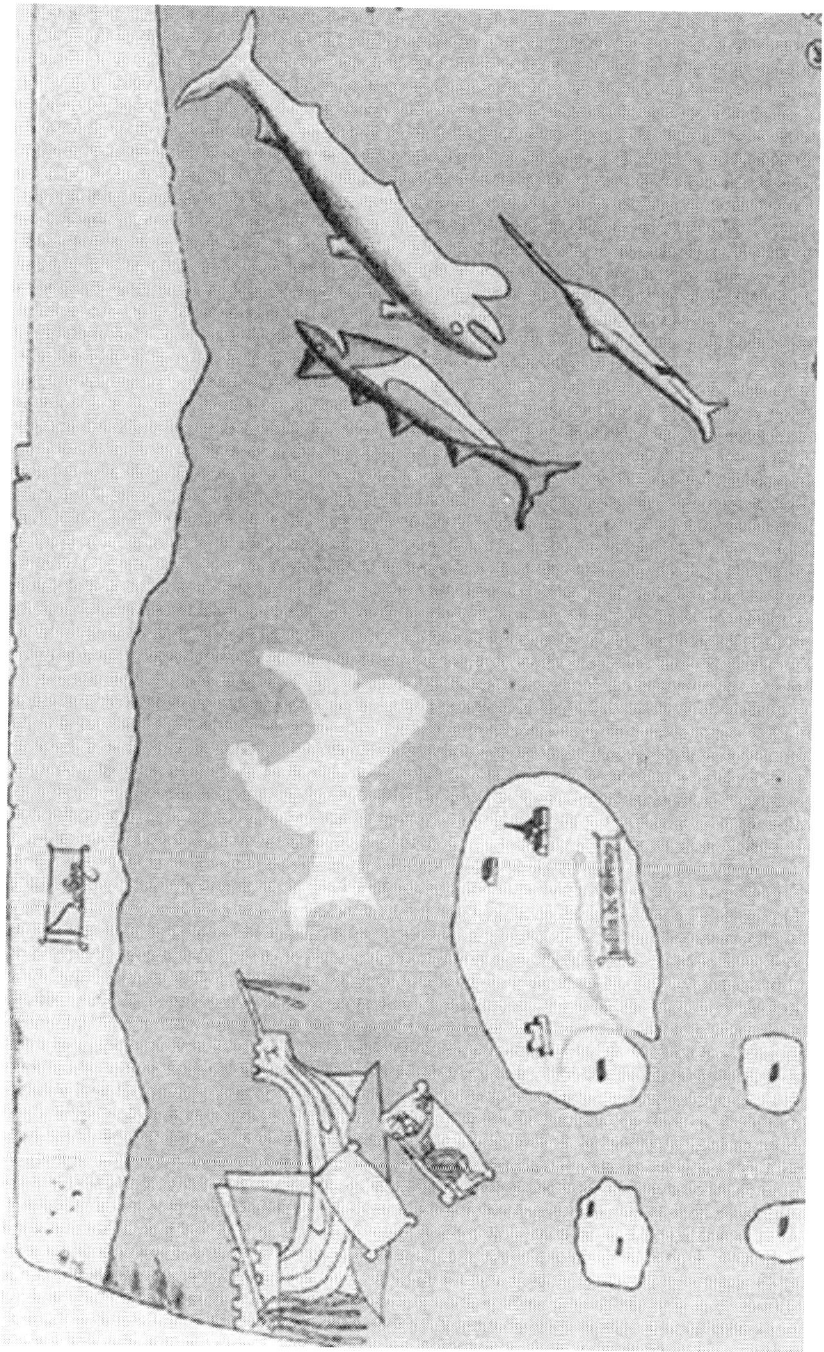

Figure 3.10. North Sea Illustrations: wrecked ship, one mariner and battle of sea-creatures; detail from the Gough Map of Britain, c.1280–1360; see fig. 3.1 for a sense of these illustrations' placement with respect to the image overall; reproduced by permission of the Bodleian Library and the Royal Geographical Society.

Figure 3.11. Henry III's Elephant, from Matthew Paris's *Liber Additamentorum*; London, B.L. MS Cotton Nero D.I, fol. 169v; c.1255; sketch apparently based on direct observation; reproduced by permission of the British Library.

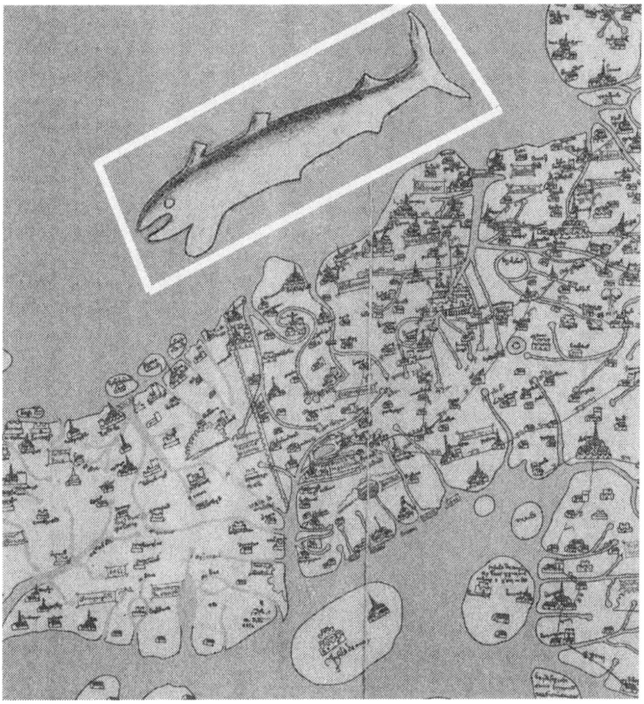

Figure 3.12. Whale and Scottish March, superimposed; detail from the Gough Map of Britain, c.1280–1360; note correspondence in spatial alignment and topography/physiognomy; Oxford, Bodleian Library Gough Gen.Top.16; RGS facsimile version, from Crone (1961); reproduced by permission of the Bodleian Library and the Royal Geographical Society.

Figure 3.13. Thrasher and Scalloped Hills of the Scottish Highlands, superimposed; detail from the Gough Map of Britain, c.1280–1360; Oxford, Bodleian Library Gough Gen.Top.16; RGS facsimile version, from Crone (1961); reproduced by permission of the Bodleian Library and the Royal Geographical Society.

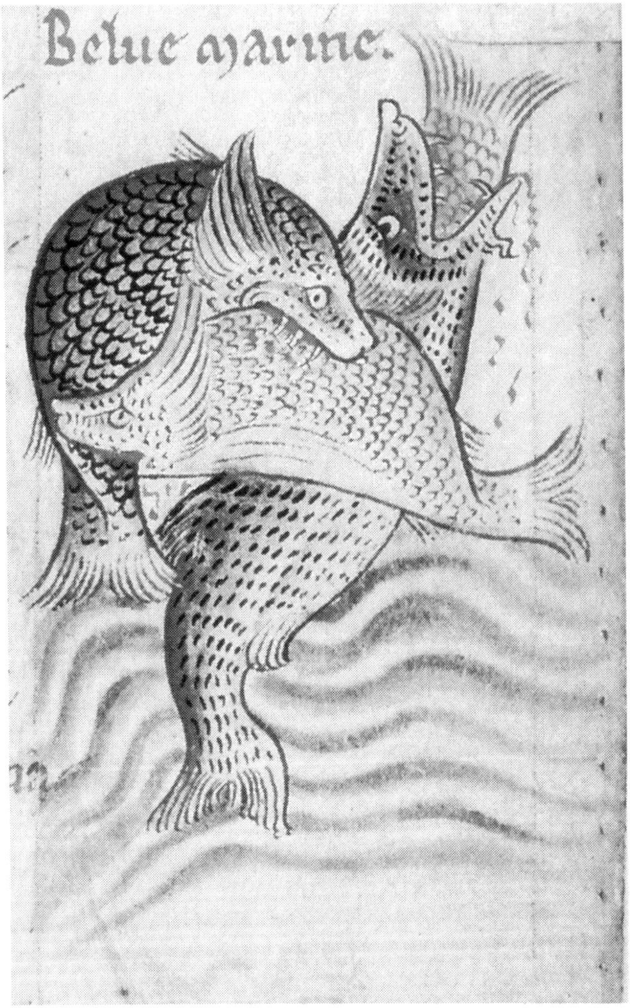

Figure 3.14. *Belue Marine* or "Sea-Monsters"; illustration from Matthew Paris's *Chronica Majora*; Cambridge, Corpus Christi College MS 16, fol. 141r; St. Albans, c.1250; reproduced by permission of the Master and Fellows of Corpus Christi College, Cambridge.

Figure 3.15: Fourteenth-Century English Coins: the Florin, with king seated on throne, issued Jan.-Aug. 1344; the Noble, with king seated in a ship, bearing sword and shield of England, issued 1344 and ff. (those minted at Calais carry "a small flag at the stern of the king's ship"—one of the earliest uses of the cross of St. George "as a national emblem"); and the Noble, temps. Edward IV; from Donald C. Baker, "Gold Coins in Mediaeval English Literature," *Speculum* 36 (1961) 282–87, Plate 1; reproduced by permission of the Medieval Academy of America.

Notes

BIBLIOGRAPHICAL NOTE

In the time that has elapsed between my initial preparation of the arguments to follow (research, conception, drafting, footnoting) and their appearance here in published form, a number of important studies on medieval cartography have either arrived in the library or been brought to my (delinquent) attention. These include (but are not limited to) Catherine Delano-Smith and Roger J. P. Kain's *English Maps: A History* (British Library, 2001) and Naomi Reed Kline's *Maps of Medieval Thought: The Hereford Paradigm* (Boydell, 2001), to cite just two outstanding examples. Other work sure to be influential remains on order (e.g., Scott Westrem's *The Hereford Map: A Transcription and Translation of the Legends with Commentary*) or is itself presently in production (for example, the proceedings to the 1999 Mappa Mundi Conference at Hereford). I have tried, through the limited means of footnote additions, to indicate where contemporary research touches against my own arguments; I regret, however, not being able to incorporate this rich fund of new material in a full and systematic sense.

NOTES TO THE PROLOGUE

[1] E. W. Tristam, *English Medieval Wall Painting: The Thirteenth Century* (Oxford, 1950), p. 180. For Wardrobe Accounts, see *Liber quotidianus Contrarotulatoris Garderobae* (Society of Antiquaries, 1789), p. 348; also Otto Lehmann-Brockhaus, *Laateinische Schriftquellen zur kunst in England, Wales und Schottland, vom Jahre 901 biz zum Jahre 1307*, 5 vols. (Munich, 1955–60): vol. III, pp. 301–2, item #6261. For the 1302 entry, which to my knowledge has never been noted by any historian of English cartography, see Lehmann-Brockhaus vol. II, pp. 222–224, #2969. For notices of the 1299 entry, see N. J. Morgan, *Early Gothic Manuscripts* II (London, 1988) p. 195; N. Denholm-Young, "The *Mappa Mundi* of Richard of Haldingham at Hereford," *Speculum* 32 (1957) 307–14, esp. 348; J. B. Harley and David Woodward, eds., *The History of Cartography* vol. I (Chicago, 1987) pp. 357, 360, 368; and Juergen Schulz, "Jacopo de' Barbari's View of Venice: Map Making, City Views, and Moralized

Geography before the Year 1500," *The Art Bulletin* 60 (1978), 425–74, esp. p. 449.

[2] It has become evident from recent work (see *Bibliographical Note*) that the hold of this position is beginning to lose some of its intensity; for its earlier hegemony, see for example Marcia Kupfer, "The Lost *Mappamundi* at Chalivoy-Milon," *Speculum* 66 (1991) pp. 540–71, esp. p. 545. At an extreme (not the case in Kupfer's measured work) this reading goes so far as to claim that medieval world maps "were not meant to be read or used" at all, "but to be interpreted like scripture": Richard Salomon, qtd. in Schulz p. 449. See also Rolf Lindemann, "A New Dating of the Ebsdorf *Mappamundi,*" in Monique Pelletier, ed., *Géographie du Monde au Moyen Âge et a la Renaissance* (Paris, 1989), pp. 45–49.

[3] See especially Paul Binski's excellent study, *The Painted Chamber at Westminster Palace* (London, 1986). Tristam is a key resource for Henry's patronage of art overall. Henry is probably remembered most fondly as a patron of art and architecture; it is often suggested or implied that his decorative projects at Westminster and elsewhere were the preoccupation of his adult life, against which the nice affairs of state were seen as a kind of untidy distraction (see final note to chapter one, below). This would be going too far. Considering the "history of Henry's troublesome dealings with his barons," Elizabeth Salter has said, "not all of it is closely related to a consideration of the cultural milieu of his court at Westminster"; see *England and International: Studies in the Literature, Art, and Patronage of Medieval England* (Cambridge, 1988) p. 76. Perhaps not, but "it would be a mistake nonetheless," as Michael Clanchy has emphasized, "to pigeonhole Henry's building work as an interest in art separated from life and politics. Through buildings and their furnishings he gave visual expression to his conception of monarchy and his own place as a king"; see *England and Its Rulers, 1066–1272* (Oxford, 1983) p. 228. Binski's work is informed by a similar approach.

[4] For detailed consideration of the Hereford Map triptych (with full references), see most recently Naomi Reed Kline, *Maps of Medieval Thought: The Hereford Paradigm* (Woodbridge, Suffolk, 2001) pp. 2–5, 76–80, 202–3—a book I encountered too late to do more than footnote, although this discussion has implications for my comments below on maps as altarpieces and/or sanctuary decoration. The literature on the Hereford Map is extensive; for starters see P. D. A. Harvey's excellent overview, *Mappa Mundi: The Hereford World Map* (London, 1996), as well as G. R. Crone, *The World Map by Richard of Haldingham in Hereford Cathedral, AD 1285* (London, 1954), a nine-part facsimile with useful (though now less than fashionable) introduction.

[5] I use the term "intervention" here as a nod to comments made by David Wallace in *Chaucerian Polity: Absolutist Lineages and Associational Forms in England and Italy* (Stanford, 1997), p. 3. What Wallace observes of medieval political treatises might well be applied (with appropriate translation and qualification) to medieval geographical and cartographic texts: these works "are typically concerned to produce not a timeless statement but one of circumstantial value, to intervene in the specific struggle of a specific secular or religious ruler against a specific enemy at a particular moment." Wallace notes that "the lexicon of political thought is of fundamental importance in providing the tropes, motifs and building blocks for such interventions"—"but each intervention is peculiarly a product of its own moment." See also, however, my comments throughout regarding the need to read medieval maps in terms of plural or multiple meanings.

[6] I cite just a few sources below, for purposes of representation, but variations on these points appear (or did so until quite recently) in much if not most work in the field. See also discussion and notes below ("A Crusader Atlas") and at the start of chapter two. On "distinct traditions" see especially Harvey's contributions in Harley and Woodward (1987), but also Woodward (1989) p. 7 and Schulz pp. 441–42. On "practical" and "symbolic" see for example Schulz p. 442 and Harley and Woodward p. 506, among others. On "single functions" see Harvey (1987) e.g., pp. 284 and 493 but cf. J. B. Harley's comments in the same volume regarding multiple functions, p. 504 and ff.; also Harley (n.d.) pp. 21, 212 and Harley and Woodward p. 3. See further Harley (1981) p. 37 on the "multivalency of cartographic meaning"; also Woodward (1989) p. 7; Kupfer (1991) p. 541; Schulz p. 448; and Blunt and Rose (1994) for similar points. On the advantages of separate analysis see Harvey (1987) p. 471.

[7] For full description and a large-size facsimile reproduction (appearing below as fig. 3.1) see E. J. S. Parsons, *The Map of Great Britain circa AD 1360 Known as the Gough Map* (Oxford, 1958). As B. P. Hindle comments of the dating, however—in a view that goes back at least to R. A. Pelham in 1933—"it is likely that [the map] was originally compiled in about 1280, the extant copy being a later revision," *Maps for Local History* (London, 1988), p. 13; see also G. R. Crone, *Maps and Their Makers* (London, 1953), p. 7; see chapter two for discussion and chapter three for elaborating details. For Gough's possible links to "a road map or itinerary in the later Roman style," see Parsons p. 10. On the afterlife of the Gough Map, in particular the Beinecke "Map off Ynglonnd" (New Haven, Beinecke Library MS 558, c.1547–54) see Daniel Birkholz, "The Vernacular Map: Re-Charting English Literary History," *New Medieval Literatures* 6 (2003); also Birkholz, "The Gough Map Revisited," *Imago Mundi* (in preparation). On Gough's links to Edward I's body of King's Clerks, see chapters two and three, and for links of kings to *mappaemundi* see "Patronage, Part One: Royal Cartography," below. Finally, for an argument linking Edward I to the Hereford Map see Valerie I. J. Flint, "The Hereford Map: Its Author(s), Two Scenes, and a Border," *Transactions of the Royal Historical Society* (6th series) 8 (1998) 19–44.

[8] On "medieval political theology" see discussion of Kantorowicz in chapters one and two, below. The evidence (and references) for Gough's likely links to "official" government sources will accrue as we go along.

[9] On Roman maps, see Woodward (1987) p. 290; Harley and Woodward p. 507; Schulz pp. 447–48; O. A. W. Dilke, "Maps in the Service of the State: Roman Cartography to the End of the Augustan Era," in Harley and Woodward (1987); and especially Claude Nicolet, *Space, Geography, and Politics in the Early Roman Empire* (Ann Arbor, 1991). For the comment on maps and colonial history, see J. B. Harley, "Maps, Knowledge, and Power" in *The Iconography of Landscape*, eds. Denis Cosgrove and Stephen Daniels (Cambridge, 1988), p. 277. This is (to my mind) arguably the single most important essay in the field. For maps and government, especially in a postmedieval context, see David Buisseret, ed., *Monarchs, Ministers, and Maps: The Emergence of Cartography as a Tool of Government in Early Modern Europe* (Chicago, 1992).

[10] As noted above this previously entrenched interpretation shows signs of receding, among specialists at least, in favor of less singular readings—still popular among non-

medievalists invested in the idea of a monolithic belief-oriented Middle Ages, however. For the incidental use of "preaching" and "teaching" see Harvey (1996) p. 14, but also Lindemann p. 45; Schulz p. 449n; Kupfer (1991) p. 545; and Woodward (1987) p. 286 for comparable points.

11 Harley (1988) p. 300.

12 Harley (1988) p. 300; Harley and Woodward pp. 506, 508. In Harley and Woodward's view, work whose "principle concern" is "the study of the map in human terms" is to be preferred over that which approaches cartography from a "narrower perspective" and which see maps as "isolatable documents, of interest primarily for their relation to one another and to the history of a teleological development of cartographic technique"; pp. 1, xvii.

13 On elites, see Harley and Woodward, esp. pp. 506–507. For speculation on the presence of maps in religious houses, see N. J. W. Thrower, *Maps and Man* (Englewood Cliffs NJ, 1972) pp. 33–34 and more recently Peter Barber, "Old Encounters New: The Aslake World Map," and Graham Haslam, "The Duchy of Cornwall Map Fragment," both in Pelletier (1989), pp. 69–88 and 33–44.

14 For a useful overview of ecclesiastical maps see P. D. A. Harvey (1996) pp. 21–39. Though I pay it the compliment of nitpicking, Woodward's landmark survey of the genre (1987) remains indispensable. Also containing much of use (especially on the close link between maps and histories, especially universal histories) is Evelyn Edson's *Mapping Time and Space: How Medieval Mapmakers Viewed Their World* (London, 1997). For details on the Aslake Map (associated with Creake) and the Duchy of Cornwall Map (associated with Ashridge), see respectively Barber (1989) and Haslam. For Chalivoy-Milon, see Kupfer; qtd. material from p. 555. For Turin, see Ernst Kitzinger, "World Map and Fortune's Wheel: A Medieval Mosaic Floor in Turin," *Proceedings of the American Philosophical Society* 117.5 (1973) pp. 344–373. For Ebsdorf, see for brief notices Lindemann as well as Armin Wolf, "News on the Ebsdorf World Map: Date, Origins, Authorship," both in Pelletier (1989) pp. 45–49 and 51–68. For recent treatment of Hereford, with attention to related productions, see Kline (2001). For the term "church cartography" see L. Bagrow, *History of Cartography*, revised R. Skelton (Cambridge, 1964), pp. 41–50, esp. 42–43, and 72–73. For subsequent citation see for example Kupfer (1991) p. 556. The "altarpiece" suggestion was made first by Richard Gough, *British Topography* (1780) vol. 1, p. 17; it has been revived frequently, most recently by Harvey (1996) p. 14 but see also Anna-Dorothee von den Brincken (1968) p. 125; Schulz p. 449 note 78; and Meryl Jancey, *Mappa Mundi: The Map of the World at Hereford Cathedral* (Hereford, 1987) p. 4; see Kline for current discussion, as noted above. For a second early witness to the Hereford Map's possible context, see Thomas Dingley, *History from Marble*, ed. J. G. Nichols (Westminster, 1867–8).

15 On a dominant theological function see for example Harley and Woodward pp. 4, 5, 506; also Woodward (1987) p. 286; for his later caution see Woodward (1989) pp. 7–8; also Woodward (1987) p. 288 for the beginnings of this view.

16 See Woodward (1987) for an overview. On the classical inheritance of medieval cartography, see for example Edson pp. 18–35; on debts to Roman models in particular, see Crone (1953) pp. 5–6, Crone (1961) esp. p. 16 and Crone (1965) p. 456. Crone's suggestion that Hereford is "fundamentally a Roman map" goes too far; see also Harvey

(1996) pp. 21–26 and ff., 50 and ff. For a map centered on Rome, see my discussion of the Peutinger Table (figs. P.4 and 2.9). On biblical and empirical justification for centering maps on Jerusalem, see Woodward (1987) pp. 341, 336—who notes that this practice was less standard (especially early on) than has usually been acknowledged.

[17] For centering on Jerusalem, see previous note (but note that there seems to be growing pressure to rescind the very notion of this as convention: "In fact, few *mappaemundi* were obviously deliberately centered on Jerusalem" observe Delano-Smith and Kane—who comment nonetheless that this may "reflect the impact of the Crusades on western thinking," p. 38). For "navel of the world" and the world map's "totalizing" argument, see Sylvia Tomasch, "*Mappaemundi* and the Knight's Tale: The Geography of Power, the Technology of Control," in *Literature and Technology*, ed. Lance Schachterle and Mark Greenberg (Lehigh, PA, 1992) pp. 66–98; 68 and 79 for references. Kupfer (1991) p. 545 also employs the term "totalizing"; and see also David Harvey, *The Condition of Postmodernity* (Oxford, 1989) for a good account of Michel de Certeau's critique of the map as a "totalizing device"; Harvey notes that "the map is, in effect, a homogenization and reification of the rich diversity of spatial itineraries and spatial stories"; pp. 252–253. On the importance to cartography of geometric structure see Harley and Woodward p. 507; also discussion below ("Map Genre Overlap: Form"). For the argument that *mappaemundi* (including Hereford, Ebsdorf, and Psalter) "functioned not only as cartographic documents but also as family members of the larger category of cosmological *rotae*," one of whose definitive features is their "circular cosmic schemata," see Kline, "The Circle as a Conceptual Device," pp. 9–48, esp. 11–13. Kline's book includes a wealth of images relevant to these issues: see figures 8.5, 8.6, 8.8, 12.3, 12.4, 4.8, 4.11, 4.12, 4.14 *et.al.*

[18] As Delano-Smith and Kane put it, "eventually the old diagrams were hijacked into the new context," p. 38; and further on the link between T-O structure and the cross, see Woodward (1987) p. 334.

[19] See Lehmann-Brockhaus vol. V, pp. 297–300 for variations on *pannus*; precedents for *pannus . . . ante altare* occur, for example, in catalog items #1232 and #2950.

[20] On maps as commercial, see J. B. Harley, "The Evaluation of Early Maps: Towards a Methodology," *Imago Mundi* 22 (1970) 62–74; see Winchester typescript version p. 214 for quoted material. On maps' associations with elites see note and reference above.

[21] Woodward (1987) provides a useful table of known maps, pp. 359–368. For Henry's 1239 commission of a map for Winchester see *Liberate Rolls of Henry III, 1226–40*, p. 405; for notices see Tristam pp. 610, 180; Morgan vol. II p. 195; and Schulz, among others. Notices for Henry's Westminster map include Binski (1986) p. 44; Pamela Tudor-Craig, "The Painted Chamber at Westminster," *Archaeological Journal* 114 (1957) pp. 92–105, esp. 100–103; H.M. Colvin et al., *The History of the King's Works: The Middle Ages* (London, 1963) pp. 495, 477–78; and more recently Barber (1989) p. 78 and Haslam p. 41. For Matthew Paris's *mappamundi*, see Vaughan (1958) pp. 247, 241; Lewis (1987) pp. 372–76, 420–21; Edson pp. 122–24; and Delano-Smith and Kane pp. 15, 250 (note 26), who note that the "surviving portion of the *mappamundi*" amounts to "perhaps a quarter of [what must have been] the original." The inscription in question appears in a substantial block at the top of Cambridge, Corpus

Christi College MS 26, p. 284 (just above the site of "Jerusalem," or in other words resting upon the top of the "T" drawn by a very prominent Mediterranean); for a later copy of the map see London, B. L. Cotton MS Nero D.V, fol. 1v; see also discussion and notes below regarding these issues. For Henry III's attention to other decorative sites, see for example Tristam, Binski (1986), and Binski, *Westminster Abbey and the Plantagenets: Kingship and the Representation of Power* (New Haven, 1995).

[22] On links between maps and secular rulers, see Schulz, Dilke, and especially the collection edited by Barber (1992). For Einhard, see *Two Lives of Charlemagne*, ed. Lewis Thorpe (Middlesex, 1969) p. 89; also F. N. Estey, "Charlemagne's Silver Celestial Table," *Speculum* 18 (1943) pp. 112–117. On Roman precedents, see Dilke pp. 235, 205–6, 207–9, and especially Nicolet, regarding the marked "geographical orientation" of Rome under Augustus, who was himself "obsessed with space" and its representation; see notes to chapter one.

[23] For England's thirteenth-century geographical culture, see Woodward (1989) p. 8; for maps and English government, see Buisseret (1992). On the Vercelli Map and England, see Woodward (1987) p. 306, P. D. A. Harvey (1996) pp. 30–33, and Kitzinger p. 358 note 89. For discussion of Bicchieri, Vercelli, and the coronation of Henry III, see below, discussion and notes to chapter two.

[24] As Edson comments of the Ebsdorf Map, "the link with Gervase of Tilbury is tempting, but unprovable," p. 138; Kline (pp. 218, 233, 239) and Delano-Smith and Kane (p. 38) regard the possibility with more enthusiasm, assigning a good deal of credibility to Gervase's connections with both the Ebsdorf Map and Westminster (also England generally): e.g., "it is largely agreed that [Gervase's *Otia Imperialia*] . . . is closely associated with the Ebsdorf Map," Kline p. 218. For Gervase's royal connections, see also Lindemann p. 46, and for his presence at Westminster see Barber (1989) p. 78 and Bagrow p. 50; nb. also Denholm-Young's suggestion that he "probably had his headquarters at London" late in life, p. 310. On these and related points, see also Wolf pp. 53, 67, 64. On the relation between Gervase, the Ebsdorf Map, and Matthew Paris, see Lewis, esp. pp. 372–376; also p. 510 n.129. For "internationalism" see Salter overall.

[25] As noted above, this rubric appears on p. 284 of Cambridge, Corpus Christi College MS 26, where Paris copied another world map, apparently that of Waltham Abbey and/or an otherwise-unknown Robert Melkeley (it is unclear whether Melkeley's and Waltham's are the same or separate productions); see citations above as well as discussion and notes to chapter one, below. For a full Latin text, see Lewis (1987) p. 509, n.117. On the Westminster Map as prototype for Psalter, Aslake, and Cornwall, see also Barber (1989) p. 78, Haslam p. 41; also Edson, Kline, Delano-Smith and Kane, *et al.* The idea that Richard de Bello may have visited Westminster in the 1360s derives from Denholm-Young's proposition that he may have been resident in Kent or even London during the period; p. 310. See Kline for updated discussion, pp. 52–57, 232–33.

[26] The term "pious" is used by Paul Binski (1986) p. 6. Other commentators make the point with terms of a comparable valence—e.g., Clanchy (1983)'s "hagiographic," a term Binski employs as well, p. 45.

NOTES TO CHAPTER ONE

[1] For concentration of power, R. R. Davies, *Domination and Conquest: The Experience of Ireland, Scotland, and Wales, 1100–1300* (Cambridge, 1990) pp. 113, 126; see also, for example, M. T. Clanchy, *From Memory to Written Record: England, 1066–1307* (Oxford, 1993) p. 45 as well as *England and Its Rulers*, p. 210; and Chris Given-Wilson, *The Royal Household and the King's Affinity: Service, Politics, and Finance in England, 1360–1413* (New Haven, 1986) pp. 28–29, 34, 38; also Binski (1986) p. 45. On London's eventual cultural hegemony, see Ralph Hanna III, "The Patronage of Sir Thomas Berkeley," *Speculum* 64 (1989) pp. 878–916, esp. pp. 912–915; for inevitable decentralization, R. H. Hilton, *A Medieval Society: The West Midlands at the End of the Thirteenth Century* (London, 1966) p. 219; and for London's build to preeminence see for example M. T. Clanchy, "England in the Thirteenth Century: Power and Knowledge," in *England in the Thirteenth Century: Proceedings of the 1984 Harlaxton Symposium* , ed. W. M. Ormrod (Woodbridge, Suffolk, 1986) p. 11; also Clanchy (1983) p. 210 and Helen Jewell, *The North-South Divide: The Origins of Northern Consciousness in England* (Manchester, 1994) p. 135. For the Waverly anecdote, see Maurice Powicke, *The Thirteenth Century, 1216–1307* (Oxford, 1952) p. 337. On the relocation of treasuries, see Clanchy (1983) pp. 163–5 and recall the presence of Edward's *pannus depictus* "in the treasuries at *Westminster*," (emph. mine), as noted above.

[2] On "expertise" and the king's court, see Clanchy (1986) p. 12. For an overview of the king's affinity, see Given-Wilson pp. 203 and ff., also p. 1; for the royal household as "heart of the kingdom" see p. 23; for the verge, see pp. 48–53.

[3] Clanchy (1986) p. 12.

[4] Given-Wilson pp. 1, 28.

[5] Binski (1986) reports that the room's "actual function" is, overall, "hard to define with precision" for the thirteenth century; this is because "the routine use of great chambers of this type was only formalized in the later Middle Ages," p. 2. But see also Thomas Beaumont James, *The Palaces of Medieval England c.1050–1550* (London, 1990) p. 73 who describes the chamber as a "private" one—somewhat problematically. On the hall and the chamber as "the two focal points of every residence of every great lord," see Given-Wilson pp. 29–30; also pp. 5–6. For a diagram showing the layout of the southern part of Westminster Palace, with location of the Painted Chamber, see Binski (1986) p. 10.

[6] Binski (1986) pp. 34, 35, 135 n.10; see also Tristam p. 91; Tudor-Craig p. 90; also T.B. James p. 75; see also Colvin p. 500.

[7] Binski (1986), esp. pp. 35–36, 38, 43–44; see also Tudor-Craig pp. 90, 103. On the importance of placing a king in proper context (and how a map might help in this) see Barber (1992) p. 26.

[8] On the necessarily "largely conjectural" nature of any "attempt to provide a context" for the original scheme, see Tudor-Craig p. 103; Binski's understandable decision to treat the 1230s scheme with considerably less attention than what he gives to the 1260s and 1290s schemes—for which much material of one sort or another *does* survive—underscores the difficulties raised by the original scheme's almost total lack of hard evidence.

[9] Binski (1986) pp. 43–45; for Henry's "cultural expedition," see Tudor-Craig p. 103; nb. also Haslam p. 41: "it is true also that Henry III in later life acquired a rather different set of artistic and decorative interests."

[10] Binski (1986) p. 36; but see also Tristam's corroborating point (complete with anecdote of a lightning strike) that the bed served frequently "as a throne or seat," p. 91; like Binski, both Colvin and N. J. Morgan describe the Great Bed as the Chamber's "pièce de résistance"; see respectively p. 127 and vol. I, p. 195.

[11] See especially Binski (1986) pp. 6, 36, 45. On Henry's sacramental conception of his kingship, see discussion below. See also Powicke p. 460; Clanchy (1983) pp. 223, 229.

[12] My point regarding Binski's disinclination to treat maps in detail is by no means meant as a criticism; his work on the Painted Chamber will remain definitive, a point amply illustrated in these notes. My confidence as to the likelihood of a map having once occupied the space on the wall behind the royal bed rests on the iconographical and architectural layout of the Painted Chamber as established by the 1260s scheme together with the evidence provided by the Psalter Map miniatures as well as by such analogs as the heraldic images set behind royal beds in figure 1.14. But see also note and discussion regarding size of the St. Edward mural, below; also there is the opinion (a decidedly minority one) that Henry's map may have been a wall-hanging rather than a mural; on this point see Denholm-Young, p. 314n., who assumes (for reasons that are unclear) that Henry's Painted Chamber *mappamundi* and Edward I's Wardrobe *pannus depictus* are one and the same production.

[13] On royal/baronial polarity, see for example Powicke, e.g., pp. 123, 134 and Clanchy (1983) p. 213; also John Neville Figgis, *The Divine Right of Kings* (New York, 1896) pp. 27–28, who notes that "the very development of a constitutional system led to a counter attempt to exalt and liberate from control the royal prerogative: 'For every assertion of national right there is a counter assertion of royal autocracy.'" For the "unpopularity of overseas expeditions" see (among others) Nigel Saul, *Knights and Esquires: The Gloucestershire Gentry in the Fourteenth Century* (Oxford, 1981) p. 40. For the argument that medieval maps typically "conveyed what contemporaries would see as a straightforward message," see Harvey (1987) p. 493 as well as discussion above. The evidence for the (lost) Westminster map's having been copied into the (lost) ordinal of Matthew Paris rests in the manuscript rubric rehearsed above (see prologue discussion and notes). However, note also Lewis (1987)'s comment in passing, p. 372, that Matthew's surviving world map sketch is a copy of both the Waltham Abbey *and* Westminster Palace maps (to me, a perplexing proposition, unless it is meant that the Westminster Map is witnessed indirectly through the Waltham/Robert Melkeley Map(s) that have apparently been copied); see also Edson pp. 122–24, who follows Lewis on this point. Uncertainty reigns supreme in all of this; I pause to dwell on the matter because an insistence that *this* sketch of Matthew's does *not itself* purport to reproduce the king's Westminster *mappamundi* (copied *elsewhere*, as it seems to me Paris indicates) is essential to any reconstruction attempt which does not depend heavily on such a sketch. In any case, see Lewis's fascinating supporting point on this topic of a relationship between Matthew's *mappamundi* and Henry's: on p. 283 of Cambridge, Corpus Christi College MS 26, there appears "an extraordinary ensemble of three large-scale heads of Christ and

the Virgin"—images which Lewis judges to have been rendered (from "large cartographic pieces visible at Westminster or even St. Albans"?) something after the manner of the pendant-anatomy of the Ebsdorf Map, which "included a head of Christ very similar in type to Matthew's drawing of the Veronica in the *Chronica Majora*"; see pp. 420–21 and ff., 376.

[14] See Lewis pp. 4–5, 206, 226. On overlaps between Matthew Paris and Henry's decorative sites and in particular the proposition that he may have designed the St. Edward murals himself, see Tudor-Craig, esp. p. 95. For overviews of Matthew's cartography including good detailed discussions of individual maps in their different manuscript versions, see Vaughan (1958) pp. 235–250 and especially Lewis (1987) pp. 321–376. For Matthew's inspiration of a local map, see Harvey (1987) pp. 469–70. For speculation that Matthew may have himself designed Henry's Painted Chamber map, although without supporting argumentation or evidence, see Tristam p. 102. For the drawing of any map as a "bold conceptual initiative," see Harvey (1987) p. 283; see also pp. 464, 484, 489 for comments in a similar vein. On the relationship between Matthew Paris's finished artwork (i.e., in the *Chronica Majora* and abbreviations) and such manuscripts as his *Liber Additamentorum* and lost Ordinal, which he may have employed as sketchbooks (as well as appendices), see notes to chapter two.

[15] In general and for citations, see Woodward (1987) pp. 286, 291–92 and Harvey (1987), esp. p. 471; but nb. Woodward's recognition of "the need to admit the imperfect or provisional nature of the conclusions drawn," given that many more maps "were lost than have come down to us." Although I seek to revisit some of the particular conclusions sponsored by *The History of Cartography*, it should be noted that the overall account there provided by no means shuts down the possibility of plural and divergent modes of cartographic meaning in the period. I look overall to qualify and in some cases extend, rather than to refute the arguments presented in this most important of all works on medieval cartography to date.

[16] See also Harvey (1987) who notes that "at least some medieval draughtsmen" may not have seen the various genres of maps "as entirely unconnected," pp. 472–73.

[17] For Vesconte, see Woodward (1987) p. 314, Harvey (1987) pp. 472–3 and Schulz p. 452n. Robert Grosseteste and Roger Bacon were other notable proponents of the view that successful reconquest of Jerusalem depended on scholarly achievement in the field of scientific geography. See Egerton pp. 12, 29–30 and Woodward (1989) p. 8; also notes below. See also Powicke pp. 229–30, on "peace among Christians as the [pre-]condition of a successful holy war"; Powicke further notes that Bacon "wrote his greatest work in the years (1266–68) when Edward I was helping to restore England to peace and preparing to join St. Louis [of France] on crusade."

[18] In general, see Woodward's overview (1987) for the various streams of medieval T-O maps (e.g., those connected with Orosius, Isidore, Beatus, Ranulph Higden). Edson also provides useful work of classification. For Ebsdorf and crusade propaganda see Wolf p. 67 as well as Bernd-Ulrich Hucker, *Kaiser Otto IV* (Hanover, 1990); on the "apostolic meaning" of Beatus maps see Schulz p. 448; also Woodward (1987) p. 357 and Tomasch p. 71; see also notes to prologue. For the ecumenical implications of Orosian geography (impacting maps like Hereford) see David Lawton, "The Surveying Subject and the 'Whole World' of Belief: Three Case Studies," *New Medieval Literatures* 4 (2000), 9–37,

esp. 19–26. On crusader embarrassment, see Samuel Y. Egerton Jr., "From Mental Matrix to *Mappamundi* to Christian Empire: The Heritage of Ptolemaic Cartography in the Renaissance," in *Art and Cartography: Six Historical Essays*, ed. David Woodward (Chicago, 1987) pp. 27 and 47, who notes that the viewer was "reminded of his Christian duty" not least by "the resemblance of this circular diagram to the Eucharist," p. 29. See also Clanchy (1983) pp. 20–22 and Daniel Williams, "Matthew Paris and the Thirteenth-Century Prospect of Asia," in *England in the Thirteenth-Century: Proceedings of the 1989 Harlaxton Symposium*, ed. W. M. Ormrod (Stamford, 1991) pp. 51–67. From the "shameful reality" that Jerusalem was held by the infidel, Williams notes, "all concepts of the known world and Christian endeavor spring. The conception of Jerusalem as the center of the world in the thirteenth century was not a superstitious misconception. It was a reality which loomed large upon the mental horizons of the population of Christendom because of its very practical, fiscal and military implications." On the other hand see qualifications of recent cartographic historians regarding our overtendency to accept Jerusalem-centeredness as a widespread map-convention.

[19] For Richard of Cornwall, see Powicke pp. 104–106; for Edward, pp. 221–24; for the term "great expedition," p. 83. For discussion of the relationship between Maps of Jerusalem and crusade, see discussion in chapter two as well as Kline, esp. pp. 211–215.

[20] See Powicke p. 106 for the baronage's lack of interest. For the transformation of Henry's crusade vow into war against the Hohenstaufen, see also Clanchy (1983) p. 238. For the connection between Richard of Cornwall, Henry's Apulian enterprise and Matthew Paris's itinerary see Vaughan (1958) pp. 239, 247–48 but especially the extended discussion of Lewis (1987), esp. pp. 325–326 and ff.

[21] See Vaughan (1958) pp. 239, 248; for references in the chronicle see Matthew Paris's *Chronica Majora*, ed. H. R. Luard, Rolls Series, vol. V pp. 346–7 and IV pp. 144–5. On Matthew's architectural sketches, Vaughan (1958) p. 249; also P. D. A. Harvey (1987) p. 495. Note incidentally that alternative routes are in fact shown for some legs of the journey between London and Apulia; however, these routes are typically much less accurately set down than is the main route (Vaughan p. 249) and if followed by the traveler would result in a longer journey (Lewis p. 342).

[22] Paris's Maps of the Holy Land appear in London, B.L. MS Royal 14.c.vii, fol. 4v-5r, and Oxford, Corpus Christi College MS 2, fol. 2v; see Vaughan (1958) pp. 245–247, Harvey (1987) pp. 495–96, and especially Lewis (1987) pp. 323–332, esp. p. 326, who sees the former not as part of a larger itinerary from London all the way to Jerusalem (Vaughan's position, which I incline to, on the basis of adjacent manuscript placement) but as a separate map, representing the "Crusader Kingdom of Acre" in its own right.

[23] On the duty of crusade, see Powicke pp. 80–81. On Simon de Montfort, see Clanchy (1983) pp. 265–66. On the "centrality of the crusade" for the royal household, see Robert Bartlett, *The Making of Europe: Conquest, Colonization, and Cultural Change* (Princeton, NJ, 1993) p. 262. On Richard of Cornwall taking the cross, Powicke pp. 104–105. On the efficacy of world maps and/or world map imagery as a substitute for crusade obligations, I am following the suggestion of Virginia Kaufmann, "Creatures from the Ends of the Earth at the Guelph Court at Braunschweig," paper delivered at the University of Minnesota, "Strangers in Medieval Society" Conference, February 1994, esp. p. 21. Kaufmann notes that in the late twelfth century Henry the Lion—Welf

Duke of Bavaria and Saxony, father of Otto IV and grandfather of Otto the Child—employed elements of *mappamundi* iconography (e.g., "creatures from the ends of the earth") in sculpture and wallpainting he commissioned for his court at Brunswick. He appears to have done so in order to characterize his conflicted relationship to Hohenstaufen empire, in particular to justify his non-participation in the Frederick Barbarossa-led Third Crusade (1189–92). Henry the Lion's concern in his art patronage for what Kaufmann terms "self-representation" recalls, on a general level, the quasi-heraldic function of later English royal maps. But *mappamundi*-drawn content provided Henry the Lion, more specifically, with a means for "coming to terms with his responsibilities as a Christian ruler." His "role as a crusader—even if he did not exactly go—was something for which Henry the Lion wanted to be remembered"; see also discussion below.

[24] The suggestion of a link between violence to the Painted Chamber and the Barons' Wars appears in T. B. James, p. 75, complete with quoted material for which, however, no citation is provided. I have not yet encountered this intriguing proposition elsewhere. The two cited characterizations of Edward I are Powicke's, although "flower of chivalry" is based in medieval usage; see pp. 227–230 for discussion of Edward's character; also p. 82.

[25] In general see John Neville Figgis, *The Divine Right of Kings* (New York, 1896), esp. p. 39, for the connection of politics and religion in the early Middle Ages. As Kantorowicz adds, "political theology" in this early-medieval period "was still hedged in by the general framework of liturgical language and theological thought, since a church independent secular 'political theology' was as yet undeveloped." For this and other points by Kantorowicz used in this paragraph, see *The King's Two Bodies: A Study in Medieval Political Theology* (Princeton, NJ, 1957) pp. 47, 59, 52.

[26] Kantorowicz, pp. 59, 52; also p. 87.

[27] See in general Powicke p. 460 and Clanchy (1983) pp. 223, 225, 229. On the challenge to Henry's influence in the localities and the *Laudus Regiae*, see Powicke pp. 209–210. On the *mal du roy* (also *morbius regius*) see *The Dictionary of the Middle Ages*, ed. J. R. Strayer (New York, 1983–85) vol. VII, pp. 255–56; see also Michael Prestwich, "The Piety of Edward I" in Ormrod (1986), pp. 120–128, esp pp. 124–26; also Kantorowicz p. 252.

[28] For speculation on the relationship between the miniature Psalter Maps and the king's lost Westminster *mappamundi*, see especially Haslam pp. 36 and 41 (esp. note 22); but also Barber (1989) p. 78; P. D. A. Harvey (1996) p. 29; Morgan vol. I, p. 195. Tudor-Craig's observation that several stylistic details of the Psalter Maps' illumination "suggest a Westminster context," p. 101n, fits well with the now prevalent view that these may well be "direct copies of the lost map from Westminster" (Haslam); cf. Edson, Kline, Delano-Smith and Kane for mentions which take this with varying degrees of faith. For full discussion see below. Note also older and recent speculation that while both the Psalter Maps and the (related) Duchy of Cornwall Map date to the mid-to-late thirteenth century, each has "archaic features" which seem to bear witness to their being copies of an older map, indeed one from c.1230s; see Jervis p. 83 and Haslam p. 41. Like Psalter, incidentally, the Duchy of Cornwall map appears to be "a court production with a London affinity," as Haslam puts it.

[29] For brief representative notices concerning Christ's body on the Psalter Map, see for example John Block Friedman, *The Monstrous Races in Medieval Art and Thought* (Cambridge, MA, 1981), p. 45 and Harvey (1996) pp. 29–30. The link to the Eucharist applies to other circular maps as well, of course, but may have more relevance for Psalter given this particular disc's size. Regarding the standard link between the Eucharist and the Incarnation, see David Lyle Jeffrey, ed., *A Dictionary of Biblical Tradition in English Literature* (Grand Rapids, MI, 1992) pp. 163–64. This entry cites V. A. Kolve's observation in *The Play Called Corpus Christi* (Stanford, CA, 1966) that "in celebrating the gift of the Eucharist, Corpus Christi also celebrates the mystical body of Christ and therefore entails the history of salvation" from Creation to Doomsday. See also Kline for discussion of the Psalter Maps in the context of the Eucharist, pp. 228–32.

[30] On winds and portolan charts, see Edward Luther Stevenson, *Portolan Charts: Their Origin and Characteristics* (New York, 1911) p. 22; Jonathan T. Lanman, *On the Origin of Portolan Charts* (Chicago, 1987) p. 49; and entry by Silvio A. Bedini in the *Dictionary of the Middle Ages* vol. 10, pp. 33–35. On scientific learning, see for example Michael Camille, "Illustrations in Harley MS 3487 and the Perception of Aristotle's *Libri naturales* in Thirteenth-Century England," in Ormrod (1986) pp. 31–44, esp. pp. 31, 37; and on winds and world maps, see Kitzinger pp. 358, 368. On the Painted Chamber's bestiary and evangelists, see Tristam pp. 92–94 and especially Binski (1986) pp. 16–17, 35, 43. For text of the commission that alludes to the bestiary see *Calendar of Close Rolls* for 1237, p. 484.

[31] Michael Camille, *Image on the Edge: The Margins of Medieval Art* (Cambridge, MA, 1992) p. 14. On Psalter's orb, see P. D. A. Harvey (1996) p. 30; for globes and orbs generally, see Harley and Woodward plates 10–11; also discussion of T-O images in Psalter iconography, below.

[32] On the apparent interchangeability of the wyvern and the dragon as well as their basic symbolic meaning, see E. Tasker and J. Beaumont, *Encyclopedia of Medieval Church Art* (London, 1993), p. 230; see also the rich entry by Marjorie Reeves in *A Dictionary of Biblical Tradition in English Literature*, pp. 210–213. For Tristam's dragon comment and for the fleur-de-lis' association with the Virgin and hence redemption, see p. 133. "*Ecce Homo*" is from the Gospel of John (19:15): "'Behold the man,' said Pilate as he presented Jesus to the Jews, crowned with thorns and wearing a purple robe."

On fleurs-de-lis, H. Gough and J. Parker, *A Glossary of Terms Used in Heraldry* (Oxford, 1894), observe that while there is "much controversy concerning the origin of the fleur-de-lis as an heraldic symbol," after its adoption by King Louis VII in the twelfth century it becomes "essentially the Royal Badge of France," pp. 266–68. Julian Franklyn and John Tanner, *An Encyclopedic Dictionary of Heraldry* (London, 1970), concur but call it "representative of [both] the French Royal Family and of France as a Sovereign State," even while they point out its "very frequent occurrence in the heraldry of all nations" and indeed suggest that "on most occasions it is used as a meaningless item of decoration." Still, Franklyn and Tanner note that it is "accepted as a symbol of the Holy Trinity," p. 136. A. C. Fox-Davies, *A Complete Guide to Heraldry*, revised by J.P. Brooke (London, 1909/1969), provides somewhat fuller treatment and notes that "the presence of fleurs-de-lis on crowns and scepters goes as far back as the fifth century in France, and that by the time of Louis VIII *Azure, seme-de-lis or* have become the fixed and determined arms of France." But if the "lily of the valley" is "typologically applied" to the Virgin Mary throughout the Middle Ages, and suggests the Incarnation as well as Creation more generally (see *Dictionary of Biblical Tradition*, entries for Incarnation and Lilies of the Valley; cf. the Sermon on the Mount, Matthew 6: 28–30, Luke 12: 27–31), note on the other hand that Fox-Davies finds the term "quite as likely to be a corruption of fleur-de-*loys*," i.e., Fleur *of Louis*, given that naming puns of this sort are common; p. 307. See also discussion below.

[33] The phrase is Kantorowicz's, p. 87: "the king, by his consecration, was bound to the altar as 'king' and not only . . . as a private person. He was 'liturgical' as a king because, and insofar as, he represented and 'imitated' the image of the living Christ."

[34] By comparison, the Ebsdorf Map covered roughly three meters square; the Hereford Map's gable-shaped skin ran 1.58 meters from apex to bottom and 1.3 meters across; and the Duchy of Cornwall Map (of which only a fragment remains) probably measured in at slightly larger than Hereford; see Harvey (1996) pp. 3 and 33; p. 29 for the size of the Psalter Map. See Binski (1986) p. 3 for the size of St. Edward's Coronation. It should be acknowledged that the precise placement of Henry's *mappamundi* within his Painted Chamber's first mural scheme cannot be established with absolute certainty. While I am convinced, on the basis of the iconographical, architectural, and ideological grounds here presented, no ultimately decisive material evidence or unambiguous documentary witness is at this point likely to turn up.

[35] On "fashioning a chivalric identity," see Lee Patterson, *Chaucer and the Subject of History* (Madison, 1991), esp. pp. 165–230, who has applied this phrase to Richard II early in his reign, p. 187. See also pp. 176–77 ("chivalric piety is a blend of sacred and secular") and p. 178 ("chivalry must be understood as the central form of self-definition by which the noble class situated itself in medieval society"). On arms characterizing the person, see O. Neubecker, *A Guide to Heraldry* (New York, 1979) p. 188. On the timetable of heraldry's emergence, see Ann Payne's entry in *Age of Chivalry* (Alexander and Binski, 1987), pp. 55–56, who notes that while the "systematic use of hereditary devices centered on the shield" has been traced back to the second quarter of the twelfth century, "by the early thirteenth century the forms and usages of heraldry had become subject to certain broadly clear conventions." On armorial bearings' figuring of genealogical privilege, see Patterson p. 181; for subsequent points, pp. 182, 168.

[36] Payne pp. 56, 57. For the 1260s Painted Chamber's Solder Guardians and their narration of virtue triumphing over vice, see Binski (1986) pp. 4–6, 13–14, 26, 32–38, and esp. 41–43.

[37] For the "Henry of Mainz" Map, also known as the Durham Cathedral Map or (most recently) the Sawley Map because it appears in a Durham copy of Honorius of Autun's *Imago Mundi* made for presentation to Sawley Abbey, see Harvey (1996) pp. 22–23 and especially "The Sawley World Map and Other World Maps in Twelfth-Century England," *Imago Mundi* 47 (1997), pp. 33–42; also Delano-Smith and Kane pp. 36–37, 12. Incidentally, once placed within its triptych, the Hereford Map too might be said to bear some resemblance to this basic heraldic look. That is, the figures which flank the map once it has been secured in the main case serve a comparable role to Sawley's supporting angels; see Harvey pp. 11–16 for more on the Hereford triptych, and especially Kline, who opens with a reconstructed figure of this artifact, p. 3. Finally, for

a compelling photograph from Westminster Abbey, of a censing angel in a doorway spandrel that resembles those on the recto Psalter Map, see Binski (1995) p. v.

For the heraldic definitions employed here, see J. Woodward and G. Bennett, *A Treatise on Heraldry* (London, 1896/1969); on punning see also Fox-Davies p. 307 and the above note. On marshalling (as well as heraldry overall) see Helmut Nickel, *Dictionary of the Middle Ages* vol. VI, pp. 172–77; nb. incidentally that in 1340 Edward III officially incorporated the Arms of France into the Shield of England to illustrate his claim to the French crown. On the proliferation of fleurs-de-lis at Westminster generally, see Tudor-Craig p. 101.

[38] See Kantorowicz pp. 419–37 for the *Lit de Justice* and the *Lit d'Honneur* and English analogs. On the crouching lion and dragon on Henry's 1218 seal, see *Age of Chivalry* (Alexander and Binski, 1987), pp. 302, 397 (such beasts are "borrowed from representations of Christ triumphant" and "are probably a reference . . . to divine support for Christian sovereignty").

[39] Neubecker pp. 188–89; Hasler p. 19; Payne p. 55.

[40] Neubecker pp. 8–9; Hasler p. 31.

[41] On badges, see Payne pp. 57–59; on the development of supporters, see Neubecker pp. 8, 189; also Payne pp. 56–57. On Matthew Paris, "an accomplished student of heraldry when armorial distinctions were only beginning to be used," see M. H. Marshall, "Thirteenth-Century Culture as Illustrated by Matthew Paris," *Speculum* 14 (1939), pp. 465–77, esp. 475–77; see also Lewis (1987) pp. 198–201, 411, 41–43, 174–76, 419, 467–68, 493 n.62, and Vaughan (1958) pp. 250–53. On the roof of Westminster Hall, see Neubecker p. 8; on seals, "particularly dragons," see Payne p. 59.

[42] Payne p. 59 for "maturity"; p. 56 on professional heralds; on rolls of arms, see references to M. Paris in previous note; on heraldry "loading the memory," Payne p. 55, who quotes Norfolk antiquary and herald Peter Le Neve (1696); p. 59 for "living art."

[43] *Age of Chivalry* (Alexander and Binski, 1987), p. 316 for changes to Henry's Great Seal; see Hasler p. 57 for leopards as English royal supports; on Henry's new throne, see Francis Wormald, *Collected Writings vol. II: Studies in English and Continental Art* (London, 1988), esp. "The Throne of Solomon and St. Edward's Chair," pp. 61–69, esp. 66–67. Incidentally, the leopard supporters were added to his throne on Henry's order, but on the recommendation of M. Edward of Westminster, the king's artistic advisor—the same man who appears to have overseen the painting of Henry's original Painted Chamber mural scheme c.1236–39. See Tristam pp. 87, 93 for discussion.

[44] For "major reorientation," see *Age of Chivalry* (Alexander and Binski, 1987), p. 316; Wormald notes that Henry was "ever interested in defining his position as king," p. 68. On the visit to Paris, see Tudor-Craig p. 104. On Henry's new enthusiasm for arms, Payne notes the opportunity provided "by a splendid set of floor tiles in the chapter house" at Westminster Abbey, laid 1253–58; see also Binski (1996) on the subject of Henry's stone armorial in the nave at Westminster Abbey, pp. 76–84; also Binski (1986) p. 43.

[45] On heraldry and the Wilton Diptych, see Wormald pp. 104, 110–114. For an example of an overturned shield in the margin of the *Chronica Majora*, see Richard Vaughan, *The Illustrated Chronicles of Matthew Paris: Observations of Thirteenth-Century Life* (Cambridge, 1993) p. 198.

[46] Going one step further, the page's foliated and triangulated outside border might also be interpreted as having heraldic relevance, on one hand in that triangles invoke the Trinity (once again), and on the other in that this foliation in the page's border complements the spandrel fleurs-de-lis in suggesting a Westminster and hence specifically English royal context. See Tudor-Craig p. 101. Henry, that is, may have strewn his Westminster decorative schemes with fleurs-de-lis in order to stress his close personal and ideological ties with the royal house of France; from 1235 he and Louis were brothers-in-law, through their wives, and from the 1250s onwards, as we have seen, he looked to Louis as a model for both his art and his kingship. The invocation of France would have been especially relevant after 1259, when Louis bailed Henry out by ruling in his favor against the English baronage in the Mise of Amiens; see Powicke pp. 182–84.

For one final link between cartographic and heraldic practice in medieval England, see Peter Barber, "The Evesham World Map: A Late Medieval English View of God and the World," *Imago Mundi* 47 (1995), 13–33, esp. 13. Something noteworthy about newly discovered Evesham, a wall map commissioned c.1390 and added to or amended approximately twenty years later (and then revised once more by 1452), is that it was "discovered in the College of Arms of London, England's most important heraldic institution, where coats of arms were officially devised and registered." Indeed the map occupies "the verso of the sixth membrane of a lengthy pedigree of eleven membranes tracing the descent of King Henry VI and of Sir Ralph Boteler" (executed 1447–1452).

[47] Among the other geographic and administrative images favored by Henry in the 1230s were a city-view of Winchester; lines of seated kings at Windsor and Rochester, each with an architectural detail in hand; and Wheels of Fortune (on whose relation to geography and cartography, see Kitzinger) also at Winchester, and at Clarendon; see Tristam and Binski (1986). For Wormald's characterization of Henry as obsessed with "defining his position," see note above. On the role of territorial maps in later periods as signs of the state, see for example Barber (1992) and Buisseret (1992) overall; also Harvey, *Maps in Tudor England* (Chicago, 1993) and Harley (1988) esp. pp. 283–84 and 300–303.

[48] See Vaughan (1993) p. ix for "famous and remarkable." On Frederick's likely influence on Henry's ideas about rulership, see Powicke p. 97; see also discussion below ("Kingship, Part Two: Legal and Administrative"). Dates for Frederick II are as follows: King of Sicily from age three (through his mother Constance, daughter of Roger II of Sicily); King of Germany from 1215—with Philip Augustus of France's defeat of Otto IV at the Battle of Bouvines being decisive; Holy Roman Emperor from 1220 (his father Frederick I Barbarossa had held both these titles); and King of Jerusalem beginning 1227/9.

[49] Kantorowicz is the champion of Frederick II's influence on Western political thought. However, as Karl Hampe says, "his significance cannot be fully understood by taking account of [his influence on political affairs] alone; the stimulus which he gave to cultural life was at least as great"; see *Germany Under the Salian and Hohenstaufen Emperors* (Cambridge, 1986) p. 269. On this cultural front, see in particular William Tronzo, ed. *Intellectual Life at the Court of Frederick II* (Washington, DC, 1994). On the vibrancy of Frederick II's court see also Powicke p. 97.

Frederick has had the tendency to call forth extremes, in the thirteenth century as well as the twentieth. *Stupor Mundi,* as Frederick was called by some of his contemporaries (including Matthew Paris), has the sense of both "terror of the world" and "wonder of the world"; but cf. Friedrich Heer's observation in *The Holy Roman Empire* (New York, 1967) that as such he was "feared, respected, much hated and probably totally unloved," p. 81. Still, as Hampe adds, "opinions about him differ violently; hatred and admiration are both still common today," p. 71. See also James Powell, "Economy and Society in the Kingdom of Sicily under Frederick II: Recent Perspectives," in Tronzo (1994) pp. 268–69.

Certainly the most controversial and influential among modern readings of Frederick II is Kantorowicz's; but as Peter Herde has noted, one problem with Kantorowicz's admittedly erudite reading of his medieval sources was that he "took most of his material at face value, rather than seeing it for what it was"—Hohenstaufen propaganda—behind which a "much more realistic, flexible and cautious" political behavior operated. See Herde, "Literary Activities of the Imperial and Papal Chanceries during the Struggle between Frederick II and the Papacy," in Tronzo (1994) pp. 227–239, esp. p. 239 n.80. In general see also David Abulafia, *Frederick II: A Medieval Emperor* (New York, 1992).

For the question of "proto-Nazi historiography," see Norman Cantor, *Inventing the Middle Ages: The Lives, Works, and Ideas of the Great Medievalists of the Twentieth Century* (New York, 1991) pp. 79–117; but note that Cantor himself has been roundly critiqued for his own reading of Kantorowicz. For Foggia and later Melfi as in turn the "geographical center" of his empire, notwithstanding Frederick's "extremely peripatetic" lifestyle, see Gary M. Radke, "The Palaces of Frederick II," in Tronzo (1994) pp. 179–88, esp. pp. 180–183. On political ideology, see Heer p. 83. On the increasing centralization and bureaucratization of the empire under Frederick, see Radke p. 181 and cf. T. C. Van Cleve, *The Emperor Frederick II of Hohenstaufen*, p. 256.

50 Overall see Heer, esp. pp. 83–84. On the centrality of papal-imperial conflict in the eyes of Matthew Paris, see Vaughan (1993) p. xiii; Frederick's coup in 1227/8 of arranging a ten year truce in Palestine and becoming King of Jerusalem, incidentally, only aggravated relations with the papacy (another excommunication). For Matthew Paris's comment on hatred between emperor and pope, see Vaughan (1993) p. 59; see also pp. 50, 18.

51 See Heer p. 6 on Frederick's vision of himself as "protector" of the church—although Alfred Haverkamp, *Medieval Germany, 1056–1273* (Oxford, 1988), points out the decidedly "harsh rule over the church which Frederick II exercised over his hereditary lands," p. 246. Or, as architectural historian Gary Radke phrases it, the emperor was "not much of a church builder," p. 183. On *regnum* and *sacerdotum* see Heer p. 76; also pp. 83–84.

52 On supremacy of papacy over empire, see Albert Brackmann, "The Beginning of the National State in Medieval Germany and the Norman Monarchies," in Geoffrey Barraclough, ed., *Medieval Germany, 911–1250: Essays by German Historians* (Oxford, 1938) p. 285; for subsequent points see Heer pp. 84–85.

53 On "spiritual resources" and "crusades to desacralize" see Heer p. 87. For the rebellion of secular lords "immediately and everywhere" against the papacy's "claim to

subject the world to theocratic rule," see Brachmann p. 286. For the "profound effect" of Frederick's resistance "on the intellectual life of the west," see Hampe pp. 251–52. For the rise of "the doctrine of the king as 'emperor in his own kingdom'" see Horst Fuhrmann, *Germany in the High Middle Ages* (Cambridge, 1986) p. 56; also Kantorowicz pp. 192, 51n, 288n (see index, p. 563) and Powicke p. 381. For key changes in the graphic representation—i.e., the mapping—of *regnum* and *ecclesia*, see chapter two, especially "Map Genre Overlap: Form."

[54] Fuhrmann p. 153; Heer p. 83; and see also Haverkamp p. 247. Not surprisingly, it was Frederick's brazen decision to crown *himself* as King of Jerusalem which most outraged the papacy and its advocates and helped earn him epithets like Antichrist; see M. Paris's comments in Vaughan (1993) p. 95. This act expressed most cogently, and provocatively, the Hohenstaufen claim to an authority conferred directly by God.

[55] Frederick's success depended also, of course, on questions of his military, economic, diplomatic, and other institutional power relative to that of the papacy and its allies. Heer, with the benefit of hindsight, has spoken of "Frederick's fundamentally hopeless situation *vis-à-vis* the papacy," pp. 83–85. On governmental practices having their own justification, see discussion below.

[56] On the cultural implications of Frederick's secular state articulation, see overall the collection edited by Tronzo, *Intellectual Life at the Court of Frederick II*; nb. also Brackmann p. 297: "The concentration of public life in the state itself and the royal government drove the clerical element into the background and stimulated the growth of new forces in every sphere of political, economic and cultural life." For discussion see below.

[57] See Claude Nicolet, *Space, Geography, and Politics in the Early Roman Empire* (Ann Arbor, 1991), whose project overall is to make us "comprehend the importance of geography for the mentality and ideology of the period"; see pp. 5, 7, 9. Despite the "ambiguity of the new political formulas" operative under Augustus, and despite the new empire's "ostentatious claim to restore ancient institutions," it is in many ways "a new monarchy—in great part a hereditary one" that is established in the aftermath of Octavian's victory at Actium. Going further, Nicolet suggests that Rome in the Augustan period may actually be deep in the process of "becoming . . . a territorial state," pp. 1–2. Cartography ties up with this process intimately because "in order to set boundaries to their empire and to claim to have reached those that were marked out, the Romans needed a certain perception of geographical space, of its dimensions and of the area they occupied"; see pp. 2–3, 9–11, 15–16. Augustan "attempts at controlling space" were many and various, but the most dramatic manifestation of the period's "new concern about space and its distribution" comes in the monumental world map whose compilation was undertaken by Marcus Vipsanius Agrippa (63–12 BCE) but was completed by the emperor himself; pp. 7–8, 95, 98–99, 101–103, 106, 113. By graphically depicting his empire's full extension over the bounds of the known world, and in the related project of presenting the account of his own achievements in explicitly geographical terms, Augustus may be said to have "founded Rome for a second time," this time, beyond *urbs*, as universal *imperium*; pp. 7, 9, 110–111, 115. Between them, Augustan geography and Agrippan cartography provided the blueprint for this foundation, of a centripetal but

strongly ecumenical Roman Empire. In citing Augustus, as he set out the contours of his own imperial project, Frederick II built on precisely this.

[58] See H. C. Friesleben, "The Origin of Portolan Charts," *Journal of Navigation* 37 (1984), pp. 194–99. However, for an overview of portolan charts and their development (with opening acknowledgment that ethnic and national sentiment has largely determined scholarly opinion as to the competing Catalan and Italian claims, and that portolan chart origin should probably "be looked for in terms of separate regional sources rather than in terms of linear descent from a single master chart or atlas"), see Tony Campbell, "Portolan Charts from the Late Thirteenth Century to 1500," in Harley and Woodward, pp. 371–463, esp. 444, 380, 388, 382, 446. See also Stevenson, pp. 2, 15–19; J. R. S. Phillips, *The Medieval Expansion of Europe* (Oxford, 1988), pp. 88, 214, 217, 220; and on Frederick II's naval power, Powell pp. 266–68.

[59] For the relevant political background, see Heer pp. 79–80, 85; Hampe p. 238; Haverkamp p. 239; Poole p. 449. This decisive foreign policy shift (marked in Frederick's marriage to Isabella) shelved "the old alliance with France" which had won him the empire in the first place in favor of a new one with England, and led in short order to an attendant reconciliation with the Plantagenets' allies since the time of Richard I, the Welfs. The possessions of this house, based primarily in the region of Saxony (north Germany) around Brunswick and Lüneberg, had dwindled considerably but by the gift of Frederick II were in 1235 substantially enlarged, and a new duchy created for Otto the Child, nephew of the deposed Otto IV. As Benjamin Arnold explains, "Otto's new status as *dux et princeps* as granted by his cousin Frederick II was . . . expected to heal the rift between their houses." See Matthew Bunson, *Encyclopedia of the Middle Ages* (New York, 1995), p. 352; also Heer p. 83. In terms of what might be called Frederick's potential deployment of cartography for political purposes, account must be taken of a striking change made to the feudal standing of Ebsdorf (and indeed much territory in Lower Saxony) at precisely the time of the map's commission or incorporation of patronal sponsorship. See Wolf (citation provided above) for how Ebsdorf appears, through its toponymy, to register just such an introduction of new feudal and territorial realities; pp. 60–64.

[60] On the shift from Christ-centered to law-centered kingship, see Kantorowicz overall; but esp. pp. 102, 90, 93, 126; also nb. his strong insistence (pp. 141, 115) that despite this shift "practically all the former values survived—but they were translated into the new secular and chiefly juristic modes of thinking and thus survived by transference in a secular setting." In other words, it is not that "the former liturgical values of kingship" suddenly ceased to exist but only that gradually "their substance grew paler."

[61] For feudalism's "disintegrating tendencies" see Barraclough p. 137, also p. 130. For "the centrifugal forces that threatened its peace and stability," see David Abulafia, "Ethnic Variety and its Implications: Frederick II's Relations with Jews and Muslims," in Tronzo (1994) pp. 221, 225. In general see also Powell, who agrees that Frederick's reign is "indeed the moment of decision. The economic implications of his possible victory [over the papacy] have never been explored but it seems certain that it would have made the road to state centralization much smoother." As with Edward I after him, "there is in [Frederick II's] fiscalism an understanding, albeit not fully developed, of the economic significance of taxation"; pp. 268–269. See Powicke pp. 220–224, 497–501, 508–509

and 528 on Edward I and taxation. On the place of cartography in the rise of England's bureaucratic monarchical state, see chapter three.

[62] On Frederick II's apparent push to develop cartography as a technology useful to the state, see notes and discussion above. On the special significance of an "at a glance" epistemology, see J. B. Harley, "Deconstructing the Map," in Trevor J. Barnes and James S. Duncan, eds., *Writing Worlds: Discourse, Text, and Metaphor in the Representation of Landscape* (New York, 1992), p. 243, as well as Harley (1981) p. 35 and Harley and Woodward p. 2.

[63] On Peter de Vinea, see Powicke p. 72; on Henry's animal menagerie, see T. B. James p. 90; also Lewis (1987) pp. 212–216. On Frederick's extremely well-regarded falconry treatise and the Latin and Arabic bestiaries he consulted in its preparation, see *Dictionary of the Middle Ages*, vol. II, pp. 242, 203 and ff. This treatise "stands as probably the finest biological work of the medieval period, a unique masterpiece still consulted for the art it instructs."

[64] See Powicke p. 97.

[65] On the potential affixation (or not) of the leaf which bears the two map miniatures into B.L. MS Additional 28681, see for example Tudor-Craig ("the page may not have been intended for a Psalter at all"), p. 101n; Morgan vol. I, p. 195; and Harvey (1996), who describes it as "an integral part" of the manuscript, pp. 29–30; also Kline p.223n. On the medieval Psalter as picture book, see Beryl Rowlan's entry in the *Dictionary of the Middle Ages* 2, p. 206, which references M. R. James on the subject (see discussion and notes below); also Morgan vol. I p. 15 and vol. II p. 13. On Books of Hours, see Morgan vol. I, esp. pp. 12–14.

[66] On prefatory sets of illuminations, see Morgan vol. I, esp. pp. 13–15, for a useful overview; p. 23 for the Map Psalter's links to other London productions; p. 82 for the Cleveland Museum piece, which itself "very probably was originally part of a Psalter." Regarding the Four Evangelists' frequent distribution around Christ in Majesty, recall their presence on the four walls of the Painted Chamber and on *mappaemundi* of the Beatus type; see Tristam pp. 94, 50–51 as well as discussion above.

Regarding the production of illuminated prefatory sequences, nb. Morgan vol. I, p. 14: "The illuminators who produced Psalters in this period seem usually to have been involved also in decorating Books of Hours, Bibles, Aristotle manuscripts, and bestiaries." Also note that after 1250 the sets of prefatory miniatures in Psalters become less "rich in narrative cycles"—"the tradition is continued but the emphasis on choice of subject moves from narrative scenes to images with more concentrated devotional meaning"; see Morgan, vol. I, p. 13.

[67] On the subject of the murals in the 1230s Painted Chamber, see initial discussion above and in general Tristam, Tudor-Craig, and especially Binski (1986). The Map Psalter appears originally to have begun with a London calendar and the map-leaf, to which was later affixed, c.1300, a set of full page illuminations, including the Adoration of the Magi (f.3v); the Nativity (f.4); Christ leaving the tomb (f.5v); the Crucifixion (f.6); the Annunciation (f.7v) and Christ in Majesty (f.8); see Edson p. 137. Folio 190v's Virgin and Child provides an example of a full page miniature interspersed in the text; see Morgan vol. II pp. 82–85, and note the parallel of a world map with an illustration of the Virgin and Child nearby in M. Paris's surviving sketch of a *mappamundi*, for

which see Lewis (1986) pp. 372–76, 420–21. The Map Psalter has fully painted and historiated initials with gold grounds at all the liturgical divisions except that the leaf with the initial to Psalm 109 is lacking—worth noting insofar as the miniature-subject consistently associated with Psalm 109 in contemporary English Psalters is Christ in Majesty, often holding a T-O orb. For the Painted Chamber's painted bestiary, see Tristam pp. 91–92. On the overlap between illuminators of Psalters and other books, see previous note. For the Rutland Psalter's marginal grotesques, see Morgan vol. II, pp. 78–82; incidentally the Rutland Psalter has a set of prefatory illuminations much like the Map Psalter's. On the Marvels of the East, see M. H. James, *Marvels of the East* (Roxburghe Club, 1929), who, though he notes the Marvels' relations with the Alexander Romances (which themselves incorporate T-O imagery), argues that there is "no organic connection" (by which he means a textual one) between the bestiary and the Marvels. For more see R. M. Lumiansky's entry on "Legends of Alexander the Great," in the *Manual of Writings in Middle English* vol. I, pp. 103–113; also Kline 165–90, esp. on links between Alexander and the Wheel of Fortune (in the context of *mappaemundi*).

68 For specifics on the Map Psalter overall see esp. Morgan vol. II pp. 82–85 and Kline pp.?? On the Rutland Psalter, London B.L. Additional 62925, see Morgan vol. II 10–14, 20, 23, and 78–82, esp. p. 80 for links to Map Psalter. See Morgan vol. II p. 82 for the Cleveland MS. See also Morgan vol. II pp. 191–192, for mention of an illumination of Christ seated, holding a globe (f.107)—the historiated initial to the first of the penitential psalms; London, B.L. MS Harley 928 (c.1280–90). At c.1280–90 this Book of Hours dates later than the period of our concern, obviously, but Morgan notes that "the palette" and "fluid swaying poses" in this book closely resemble the (added late thirteenth-century New Testament) scenes in the Map Psalter.

For another image in the vein of Fig. 1.23 ("an initial enclosing God the Father enthroned, carry an orb representing the world," with which the penitential psalms open), see Janet Backhouse "'A very old book': The Burdett Psalter Hours," in Brendan Cassidy and Rosemary Muir Wright, *Studies in the Illustration of the Psalter* (Stamford, 2000), 55–66. "Distinctly English symptoms within the manuscript's original context surely suggest that the patron could have been an Englishman"; Backhouse also suggests indirect connections to Edward I; "French made," probably 1380s, but "can be placed quite firmly in England" by the end of the century; pp. 64, 59. Finally, for another striking image of God enthroned (with tenant angels), holding compass in one hand and disc of the created world in the other, see Oxford, Bodleian Library MS 270b, f.iv: The Creation (God as Architect), from a thirteenth-century French *Bible moralisée*; see Kline p. 235 for reproduction.

69 See Powicke pp. 103–104 for the baronial response first to Henry's Poitevin expedition and later to the Sicilian Business. On the increasing prominence of St. Edward in Henry's decorative schemes, see for example Binski (1995) pp. 13 and ff. and Colvin p. 127; also, Christopher Wilson in *Age of Chivalry* (Alexander and Binski, 1987) p. 77, who notes that "Henry's determination to revive it [i.e., the cult of St. Edward, with its 'ancient idea of a priest-like ruler'] had already aroused stiff opposition before 1245 and was to be one of the causes of the baronial revolt of 1258–65."

70 Tudor-Craig pp. 100–101; Dilke pp. 244, 234. Of the *Notitia*'s picture maps, Dilke notes that many are "such as only a bureaucrat unfamiliar with the areas could

have produced." He also notes however that the shortcomings of the maps "do not mean that the *Notitia Dignitatum* was a useless document, but in fact that useful information would have had to be gathered from the text rather than from the maps," p. 245. In general see Roger Goodburn and Philip Bartholomew, eds., *Aspects of the Notitia Dignitatum* (Oxford, 1976); pp. 8, 81 for dating. Each of the *Notitia* essays referenced in the next few notes appears in this collection. For an overview especially useful on the *Notitia*'s relationship to visual culture, see J. J. G. Alexander, "The Illustrated Manuscripts of the *Notitia Dignitatum*," pp. 11–25, esp. p. 19 for material cited. See also earlier discussion of Augustan Rome's "geographical orientation."

[71] See M. W. C. Hassall, "Britain and the Notitia," esp. pp. 103–105, for the *Notitia* as a "prime source" for Roman Britain; as a report on the Empire's "condition and resources"; and for a detailed breakdown of its layout and purpose. Hassall turns to the *Notitia Dignitatum* in an attempt to ascertain "something about the workings of the late imperial bureaucracy as a whole." On civil as well as military, see R. S. O. Tomlin, "*Notitia Dignitatum Omnium, Tam Civilium Quam Militarium*," pp. 189–209, esp. p. 195.

[72] On the *Notitia Dignitatum* and the Carolingians, see discussion by Alexander, esp. p. 19. On Charlemagne's silver table, see Estey; also discussion above.

[73] For the tangled questions surrounding the medieval provenance of the *Notitia*, see the spirited argument of C. E. Stevens, "The *Notitia Dignitatum* in England," pp. 211–224, esp. p. 219. On Bede and Lullus, see Alexander p. 19; also M. H. James, *Marvels* p. vii. For more on Alcuin, see *Two Lives of Charlemagne*; Einhard notes that "under him the Emperor spent much time and effort in studying rhetoric, dialectic, and especially astrology. He applied himself to mathematics and traced the course of the stars with great attention and care," p. 79. Notker the Stammerer adds that "the emperor went so far as to have himself called Alcuin's pupil, and to call Alcuin his master," pp. 94–95.

[74] MS Cotton Tiberius B.V as a "medieval parallel" to the *Notitia Dignitatum* is Tudor-Craig's excellent proposition, pp. 100–101. On the manuscript's royal and London links, see M.H. James pp. viii and 6; it once belonged to Battle Abbey. For the Marvels of the East, often connected with *mappaemundi*, see above notes. Regarding the Cotton/Anglo-Saxon Map, see for example Harvey (1996), who notes that it is "closer in its outline than any other surviving medieval map to a Roman map of the world," pp. 26–27. Matthew Paris's sketch of the Waltham Abbey map has something of the Cotton/Anglo-Saxon Map's squareness. For a more complete list of the contents of Cotton Tiberius B.V, see Tudor-Craig p. 101, who like Stevens believes (on the grounds of this witness) that a copy of the *Notitia Dignitatum* now lost may have been present in thirteenth-century England. Clanchy (1993) describes Gervase of Canterbury's *Mappa Mundi* as a "gazetteer of religious houses" and notes its three-column layout, p. 133. See also Dom. David Knowles, "The *Mappa Mundi* of Gervase of Canterbury," *The Downside Review* 48 pp. 237–47. Haslam p. 43 nominates Gervase of Canterbury as a possible candidate for the compilation of the Duchy of Cornwall Map. Gervase has also received attention from scholars interested in challenging firm distinctions between textual and visual *mappaemundi* (as well as medieval geographical productions generally).

[75] On quasi-classical iconography, Westminster, and the St. Albans scriptorium see Tudor-Craig p. 102; but note Morgan's caution on the subject of a St. Albans "school,"

vol. II pp. 22–23 and esp. 12. See also Vaughan (1958) pp. 205–34 and Binski (1986) pp. 105–112, on "court style."

[76] On Windsor's line of kings and Matthew Paris, see Binski (1995) p. 58 and Binski (1986) pp. 52–53, as well as Tudor-Craig p. 102. On the Tree of Jesse, see for example Tristam p. 60, who notes that "such genealogical trees may perhaps be regarded as prototypes of the family tree which in later times was often represented above a fireplace" (which is exactly where the Painted Chamber's Jesse Tree was painted). It may be worth noting that in 1269, not long after the Painted Chamber world map scheme was painted over, a Jesse Tree and Wheel of Fortune which had been part of a mural cycle at Clarendon were likewise replaced, having lasted only 22 years. See Tristam pp. 65–79, 208–219. Note finally regarding Jesse Trees that many include a representation of God or Christ holding orb or T-O map; see Morgan for examples.

[77] Harley (1970) p. 212; see also mentions above and below.

[78] For the popular (although generally outmoded) view of medieval world maps as primarily repositories of the "fantastical" see for example Gabriel Alington's *The Hereford Mappamundi* (Hereford, 1996), esp. chapter one. See also Sotheby's focus on the map's "picture-book" qualities and status as "probably the single most important tourist attraction in the West Country" in its advertising pamphlet when the map was briefly put up for sale: see "The Hereford World Map" by Christopher de Hamel (August 1990). For similar (usually older) characterizations, see for example Jervis pp. 74–75, 85, who regards the Hereford Map as "a fantasy"; Beazley vol. III, p. 528; and Schulz p. 463, who stresses "not the realism of some [medieval world maps], but the sheer fantasy of most." Like many non-medievalists even in recent years, in briefly summarizing the geographical thought of the period David Harvey has chosen to highlight "the sinister figures of myth and imagination" which inhabit the edges of *mappaemundi,*" seeing this as "evidence of "medieval parochialism and superstition"; see *The Condition of Postmodernity* (Oxford, 1989) pp. 240–41.

[79] On document layout, see in general Clanchy (1993), pp. 132–35; Harvey (1996) has called attention to the verso map's "list of provinces and cities in the space for each continent," but argues that their meaning remains a didactic spiritual one; p. 30. For treatment of "List Maps" as an identifiable medieval sub-genre, see Edson, pp. 5–6, 14, 90, 133, 137.

[80] See Malcolm Parkes, "The Influence of the Concepts of *Ordinatio* and *Compilatio* on the Development of the Book," esp. pp. 123–26, in *Medieval Learning and Literature*, ed. J. Alexander and M. Gibson, (Oxford, 1976) pp. 115–141. For more on *compilatio*, see A. J. Minnis, "Late-Medieval Discussions of *Compilatio* and the Role of the *Compilator*," in *Beitrage zur Geschichte der Deutschen Sprache und Litteratur* 101 (1979) pp. 385–421; Ralph Hanna III, "*Compilatio* and the Wife of Bath: Latin Backgrounds, Ricardan Texts," in *Latin and Vernacular*, ed. A. J. Minnis (Cambridge, 1987) pp. 1–11; and Mary A. Rouse and Richard H. Rouse, *Authentic Witnesses: Approaches to Medieval Texts and Manuscripts* (Notre Dame, IN, 1991).

[81] Parkes, pp. 128–29.

[82] Parkes, p. 131.

[83] Suzanne Lewis, *Reading Images: Narrative Discourse and Reception in the Thirteenth-Century Illuminated Apocalypse* (New York, 1995) p. 2; Hanna (1987) p. 6

[84] For well-regarded work on the relationship between text and image, in terms of the production and visualization of *mappaemundi*, see the work of Patrick Gautier Dalché as well as Marcia Kupfer (1994).

[85] See Tudor-Craig, p. 102

[86] Morgan questions the usefulness of the concept of a "court school" and a "St. Albans school" (see note above), but see also Tudor-Craig p. 104

[87] Commentators including Binski (1986; 1995), Clanchy (1983), and Powicke present King Henry in his last years as more or less obsessed with decorative activity centering on his promotion of the royal cult. Left-handed compliments abound: Powicke, for example, describes Henry's elaborate translation of St. Edward's remains to their new shrine at Westminster as "his greatest triumph, the fulfillment of his dearest hope," p. 244. See the similar valence of Clanchy's comments, p. 227 (e.g., "greatest memorial"). No less important than St. Edward's translation, in the eyes of many, were Henry's plans for his own eventual burial in the Confessor's vacated grave at Westminster; in this way a most immediate blending of the body of Henry with that of his sainted ancestor might be secured; see Binski (1986) p. 36; also Clanchy (1983) p. 281.

NOTES TO CHAPTER TWO

[1] Sarah Beckwith, *Christ's Body: Religious Culture and Late Medieval Piety* (New York, 1993) pp. 27, 42, 45.

[2] Ernst Kantorowicz, *The King's Two Bodies: A Study in Medieval Political Theology.* (Princeton, NJ, 1957) pp. 19, 195, 197, 207, 218.

[3] Historians of cartography may be taken aback by the confidence with which I adopt (the early end of) this date in my analysis of the Gough Map, but I should stress that I do not substantially depart in this from the conclusions of previous and more authoritative scholars of the map's textuality and compilation (judgments based on a combination of telling detail and gut feeling: never to be taken lightly in an expert!); for references see notes to prologue. The arguments in support of such a dating, not easily reducible to a footnote, are set out primarily in chapter three. For references to previous scholarship and further discussion see notes below. It also bears observing that ultimately there may be Roman basis for elements of Gough's geography, while in the opposite direction maps based on Gough's model were being produced (in both manuscript and print form) as late as the second half of the sixteenth century.

[4] On links to royal patronage, see discussion in prologue. See discussion and notes below ("Map Genre Overlap: Form") for "practical/symbolic"; for Gough as a "Road Map" see W. W. Jervis, *The World in Maps: A Study in Map Evolution* (London, 1936), p. 84; also E. J. S. Parsons, *The Map of Great Britain c. AD 1360 known as the Gough Map* (Oxford, 1959) p. 7. The phrase "cartographic revolution" is G. R. Crone's; see *Early Maps of the British Isles, AD 1000–AD 1579* (London, 1961), p. 7. See also Parsons p. 4, who notes the contemporaneity of the Gough Map and (for example) the Hereford *mappamundi* but adds that "in conception" they are "years apart"; Parsons finds Gough "remarkable for its isolated position in the cartographic record." On the need to resist producing analysis whose focus is on the teleological development of cartographic form, see comments above.

[5] Kantorowicz, pp. 173–180, esp. 175; for *geminatio*/duality, see pp. 43–44. On the public sphere, see note below.

[6] Kantorowicz, pp. 189; 173–80; 43–44, also 56n. Kantorowicz's demarcation of this extra-feudal realm's outlines is worth rehearsing: "The line of distinction therefore was to be drawn between matters affecting the king alone in his relations to individual subjects, and matters affecting all subjects, that is, the whole polity, the community of the realm. . . . Better than distinguishing between the king as a private person and the king as a non-private person would be to distinguish between a king feudal and a king fiscal, provided that we mean by feudal predominantly matters touching individual relations between liegelord and vassals, and by fiscal matters touching all." This last point above all was crucial and made clear: that "the continuity of [the fisc], beyond the life of an individual king, had become a matter of common and public interest because the continuity and integrity of that domain were matters 'that touched all.'"

[7] Following the jurist Henry Bracton (fl. c.1260), whose interests were more for administrative matters than for "metaphysical ideas and theological distinctions," Kantorowicz notes that the fisc was comprised of "'those things which make the crown what it is'" and which pertain to "'common utility'"; p. 164. For "peace and justice," p. 173; for "utilities," p. 344; on "impersonal" and "quasi-sacred," pp. 177–78, 173, 184. See chapter three for detailed discussion of the fisc's various elements in regards to the Gough Map of Britain. For Wales, see especially R. R. Davies, *Domination and Conquest: The Experience of Ireland, Scotland and Wales, 1100–1300* (Cambridge, 1990).

[8] Kantorowicz, p. 173; p. 191. Kantorowicz argues for "a complete reversal of the former ideas of Christ-like or Christ-centered kingship" p. 319, but frequently stresses nonetheless that in the later period the ideas of the earlier did not necessarily lose their force; kingship remained importantly sacramental and liturgical even as it took on newly operative aspects, relating to law and territory.

[9] For "hallowing" and "bodified" see Kantorowicz p. 186; also p. 271. For *patria* see pp. 247, 238, 234; also 271–284. To put things another way, "the nascent sovereignty of the secular state" was intimately tied up with not only a general "sanctification" of such a polity, but also with the period's "budding fiscal theology": pp. 189, 194, 186.

[10] Juergen Schulz, "Jacopo de' Barbari's View of Venice: Map Making, City Views, and Moralized Geography before the year 1500," *The Art Bulletin* 60 (1978), pp. 425–74, esp. p. 442; Harley and Woodward, e.g., pp. 291–92, 507. Unlike some who insist that the various genres of medieval maps existed in essentially separate traditions, with negligible formal and functional overlap, Schulz allows that "cartographically" [i.e., formally] the groups are "interrelated"; but in line with most others he stresses that "in function they differ clearly," p. 442. This terminological binarism may well be impossible to avoid.

[11] On "Map Genre Overlap: Function" (including references) see chapter one. For the argument that medieval maps fall into "several quite distinct traditions" which "form mainly discrete groups," see especially P. D. A. Harvey, who observes "it is seldom that we find contact between them." A related point is that the separate traditions "each served a distinct purpose": "there was no such thing as a general map designed to be put to a wide variety of uses. Any one map was drawn with one particular purpose, even one particular occasion, in mind"; see Harvey's chapters in Harley and Woodward, eds., *The*

History of Cartography vol. I (Chicago, 1987), esp. pp. 283–85. The editors themselves, though at times they incline toward this view—e.g., in noting a "growing recognition that [particular] maps filled particular functions"—nonetheless also take pains to highlight the "multivalency of cartographic meaning"; see pp. 3, 504 and ff. as well as Harley (n.d.) p. 212 and further references on this point in notes to prologue, above.

[12] For a reproduction of the Brussels Map of Jerusalem as well as several others, including a useful schematic diagram, see Daniel Bahat, *The Illustrated Atlas of Jerusalem* (New York, 1990). Kline (2001) also includes a reproduction and discussion of this map, which I encountered too late to incorporate here.

[13] On crusader manuscripts and *sitii Hierusalem*, see Kenneth Nebenzahl, *Maps of the Holy Land: Images of Terra Sancta through Two Millennia* (New York, 1986), p. 32; on the Hereford Map's compass grounding, see P. D. A. Harvey, *Mappa Mundi: The Hereford World Map* (London, 1996) p. 10. On the importance of geometric structure see Harley and Woodward: "The distinctive geometric structures of maps" are nothing less than "crucial in the history of cartography," p. 505.

[14] For the representation of actual travel routes on, for example, the Hereford Map, see the well-known argument of G. R. Crone, outlined in *The World Map by Richard of Haldingham in Hereford Cathedral, AD 1285* (London, 1954), pp. 18–22, and extended in "New Light on the Hereford Map," *Geographical Journal* 131 (1965) pp. 447–62; but note also Woodward (1987)'s equally forceful response: "Although Crone drew attention to what he considered to be the route-planning function of some world maps . . . no amount of twentieth-century historiographical ingenuity can counteract the overwhelming evidence that the function of *mappaemundi* was primarily didactic and moralizing, and lay not in the communication of geographic facts," p. 342; see also p. 288. It does not seem to me that an either/or choice needs to be made; indeed the whole point would seem to be a collapsing together of "practical" *and* "symbolic" modes.

[15] See especially the discussions of Matthew Paris's Itinerary from London to Apulia (and its potential extension to Jerusalem or Acre) by Suzanne Lewis, *The Art of Matthew Paris in the Chronica Majora* (London, 1987), pp. 323–346, esp. 338, and Richard Vaughan, *Matthew Paris* (Cambridge, 1958), pp. 47–48; especially relevant is the apparent or relative "breakdown" of the itinerary once Apulia is reached, into something "looser and more maplike"; nb. also the question of Matthew's adjacent Map of Palestine as essentially an itinerary to Jerusalem (so Vaughan, who follows Konrad Miller in this) or, rather, as a map of the Crusader Kingdom of Acre (so Lewis, pp. 324–26).

[16] For the (often dizzying) set of relationships and potential relationships between the various versions of Matthew's Maps of Britain (and London-Apulia Itineraries) see esp. J. B. Mitchell, "Early Maps of Great Britain I: The Matthew Paris Maps," *The Geographical Journal* LXXXI (1933), pp. 27–34, esp. p. 32; Crone (1961), pp. 14–15; and Lewis (1987) p. 364; nb. Lewis's caveat that "although the maps are closely linked both to each other and to the texts of the Chronicles, no two are identical," p. 321. For reproductions of the Britain maps, see *Four Maps of Great Britain by Matthew Paris about AD 1250*, ed. J. P. Gilson (London, 1928). On the London-Apulia Itinerary and its versions, see discussion and notes above, but most of all Lewis (1987) pp. 323–24 and ff.

On the map as "simply a model or theory about real world structures" or "a theory which geographers had accepted" (attributed to Ullman), see David Harvey,

Explanation in Geography (London, 1969), pp. 369, 372, 373. For a more recent formulation, see for example David Harvey, *The Condition of Postmodernity: An Enquiry into the Conditions of Cultural Change* (Oxford, 1989) p. 233: "The mapping of the world opened up a new way to look upon space as open to appropriation for private uses"; note also Harvey's observation that "the ability to influence the production of space is an important means to augment social power." See also the work of J. B. Harley, especially "Maps, Knowledge, and Power" (1988) and "Deconstructing the Map" (1992).

[17] The bits of topography that do appear tend to do so as hurdles to the route, so to speak; as signs of possible navigational/wayfaring difficulty that may require negotiation. Thus at Dover we encounter water and ships; later, rivers to be crossed are marked; and eventually, in the Alps, each city or way-station has its architectural sketch "situated atop stylized hills or mountains." Susa's stork, Lewis notes, is a "harbinger of spring taken from the medieval bestiary" and is "perhaps quoted here as an emblem of safe crossing out of the treacherous alpine pass," p. 388.

[18] Paris's comment appears in MS Royal 14.c.vii, fol. 5v (see fig. 2.6); see also Lewis p. 365; Britain's "backbone" route from Dover to Newcastle runs by way of London, St. Albans, Northampton, Leicester, Newark, Pontefract, and Durham. See Crone (1961) p. 14–15, who notes incidentally that "the maps contain no fewer than 252 geographical place-names."

[19] Suggestive of just such a sense of "a view from a single route" is the fact that, as Lewis (1987) p. 367 for example notes, the "remaining towns, rivers and castles" on the Paris Map of Britain—i.e., those not on the Newcastle to Dover axis—"are placed in relation to it [i.e., the axial route] rather than to one another." Overall Lewis provides a useful and detailed description but relatively little in the way of formal interpretation of Matthew's Maps of Britain, other than to note that of the names appearing on the maps, "nearly all . . . are mentioned in the *Chronica Majora,*" p. 367. Thus we arrive at essentially the same conclusion regarding these maps as for the London-Apulia Itinerary—that this is once again simply an *estoire* or pendant to written text, an image which like the Itinerary acts as "an elaborate pictorial illustration for the historical text"; see pp. 364, 322. The limitation of this reading, reasonable as far as it goes, is that it does not take into account the significant geometric or structural differences between the map-sets—it does not come to grips with them as cartographic documents of particular and *divergent* sorts.

[20] Michel de Certeau, *The Practice of Everyday Life* (Berkeley, 1984), pp. 116–117; Crone (1954) esp. pp. 18–22, 7–9. For more on Rome and its grid-systems, see Samuel Y. Egerton Jr., "From Mental Matrix to *Mappamundi* to Christian Empire: The Heritage of Ptolemaic Cartography in the Renaissance," in *Art and Cartography: Six Historical Essays*, ed. David Woodward (Chicago, 1987), pp. 10–50, esp. p. 15 and ff. See also discussion below ("The Ideology of a Road Map").

[21] Jervis (1936) p. 83.

[22] Nicolet (1991) pp. 8–10. Rather than being an integrated single image, this map or atlas (both are probably better terms than "table," the traditional rendering) is actually an extensive series, in a roll whose top sheet (Britain) is missing, of strip maps charting itineraries in the provinces. Much as the *Cursus Honorem* did for administrative prac-

tice (this was the corps of municipal provincial magistrates for whom an elaborate station-point system of travel and mail-delivery was established under Augustus and extended in the later empire), the *Tabula Peutingaria* is designed to cover the whole empire and link it together visually. Its routes radiate from, or converge ultimately upon, a circle which contains personified Rome, which acts as linchpin in this geographic system just as Jerusalem does for medieval *mappaemundi*. The travel-system designed for the *Cursus Honorem* (literally, *cursus* denotes a running or hastening, a course or direction, or a journey by land or sea; *honorem* indicates a public honor, official dignity, office, post, or preferment) thus provided the geographic basis and graphic template for the empire's most important surviving cartographic production; see Nicolet for discussion.

[23] See in particular Vaughan (1958) pp. 65–91, esp. 81, 84–85, for the *Liber Additamentorum*. Vaughan has discussed in great detail the complicated issue of the timing of the *Liber Additamentorum*'s compilation of its various, heterogeneous documents, although he does not directly treat the question of a precise dating for the *Scema Britannie* (as this comes on one of the "odd leaves which have only fragmentary material"). Lewis (1987) provides a useful list of the contents of the *Liber Additamentorum* in her Appendix II, pp. 467–468; see note 35, p. 475 for discussion; see also N. J. Morgan, *Early Gothic Manuscripts* vol. I (London, 1988), pp. 134–36. For an edition of the *Liber Additamentorum*, see vol. VI of the *Chronica Majora* Rolls Series (1882), ed. H. R. Luard; see p. 521 for mention of the *Scema*.

On Matthew Paris's apparent practice of employing a sketch book or model book and working from this in producing the more formal and elaborate work which appears in the various versions of his chronicles, see Lewis, esp. pp. 31–32; also p. 479, n.112. To Lewis the evidence "strongly suggest[s] that Matthew kept a portfolio in which he collected drawings of figures and motifs from various sources for later use"; his single surviving *mappamundi* is, Lewis notes, one example of a piece "drawn in a careless sketchy style," p. 372. Lewis observes that "most medieval model books date from the fourteenth and fifteenth century but enough exists of earlier ones"—such as one by Villard de Honnecourt, 1230–40—"to enable us to recognize their most salient characteristics in Matthew's illustrations"; "all Matthew's illustrations in his historical works betray the pattern book's most characteristic quality of a highly finished, carefully wrought drawing as opposed to the spontaneous sketch."

[24] For dismissive treatment of the *Scema* see for example Vaughan, who observes that this sketch "need not detain us long," p. 244. M. H. Marshall p. 247 follows the lead of Edwin Guest, "The Four Roman Ways," *The Archaeological Journal* 14 (June 1857), pp. 99–118, p. 109, who in turn follows Roger Gale, "An Essay Towards the Recovery of the Courses of the Four Great Roman Ways," in *The Itinerary of John Leland the Antiquary*, vol. 6, pp. 116–150 (Oxford, 1770), in using the term "rude"; so too B. P. Hindle, *Maps for Local History* (London, 1988) p. 13, who also complains that "this map contains half a dozen fundamental errors"; others such as P. D. A. Harvey (1987) and Suzanne Lewis (1987) decline to treat it in otherwise comprehensive studies, although Lewis does provide a transcription of its rubric in her list of the *Liber Additamentorum*'s contents. For a facsimile, see Crone (1961) Plate 5; for a schematic rendering and transcription, see Gale p. 119. See also R. Gough, *British Topography* (1780) pl. 1–2, and Gale p. 96 in Leland's *Itinerary* for engravings. In contrast to most earlier surveys of English cartogra-

phy, Delano-Smith and Kane (2001) give space and insightful attention to the *Scema*—their reading of its construction in terms of the "Four Ways" chronicle tradition aligns extremely well with my own take (developed independently, I should add), as set out below.

For discussion and a reproduction of another historical/cartographic diagram sketched and written/annotated by Matthew Paris—his circular "floral diagram of the Heptarchy," in which England's seven former Anglo-Saxon kingdoms are arranged "as close as possible [with regard to their] appropriate cardinal directions"—see Edson pp. 123, 125; also Delano-Smith and Kane p. 15. This diagram survives in several versions, and though it is circular and "highly abstract," Matthew notes that "actually the shape of England is oblong, as is shown in the following pages" (see Edson).

25 On the *Scema* as oval, note incidentally the relatively elaborate oval plan or view of London in fig. 2.5, Matthew's Itinerary from London to Apulia; on the text along Fosse Way, see discussion below.

26 On the *Scema*'s status as a "spontaneous sketch" as opposed to a "highly finished, carefully wrought drawing," see the discussion in notes above. For the outline of Matthew's Britain as "demonstrably akin" to the shape of the island on (for example) the Hereford Map, see P. D. A. Harvey (1996), p. 36.

27 I follow Gale p. 119 as well as Lewis p. 468 and Luard vol. VI, p. 521 in this transcription; the translation is mine.

28 Crone (1961) p. 15; see also Lewis (1987) pp. 374–75, who is among a number of others to have made this point.

29 For Belinus's "entire island" see Geoffrey of Monmouth III.5; on Alfred, Gildas et al. see J. S. P. Tatlock, *The Legendary History of Britain: Geoffrey of Monmouth's Historia Regum Britanniae and its Early Vernacular Versions* (Berkeley, 1950), pp. 279–80; also 3–4. Tatlock notes incidentally that "all this about Gildas and Alfred, needless to say, so far as any evidence goes is sheerest moonshine; no evidence exists for a fact or tradition that either of them translated British Laws."

30 For discussion of the appearance of the trope of the Four Ways in Ranulph Higden and Henry of Huntingdon (Book I, p. 299), see Guest pp. 100–103 and 103–15, respectively; see also Gale p. 117 and ff. For Langtoft, see *The Chronicle of Pierre de Langtoft*, ed. T. Wright, Rolls Series vol. 47 (London, 1866) pp. 44–45. For Walter of Coventry, see *Memoriale Fratris Walteris de Coventria*, ed. W. Stubbs, Rolls Series vol. 58:1 (London, 1872) pp. 5–6. For Geoffrey of Monmouth (III.5 in standard version), see also *Geoffrey of Monmouth's Historia Regum Britanniae: A Variant Version*, ed. Jacob Hammer (Cambridge, MA, 1951) pp. 54–57. For Roger de Hoveden, see *Annals of Roger de Hoveden, A.D. 1180*, ed. and tr. H. T. Riley (London, 1853) pp. 545–46. For Robert of Gloucester see Gale p. 131 and ff. For Matthew Paris and Roger of Wendover, see *Flores Historiarum*, ed H. R. Luard, Rolls Series vol. 95: pp. 1–3 (London, 1890); also Matthew of Westminster, *Flowers of History*, 2 vols., ed. and tr. C. D. Yonge (London, 1853), where there is no mention of the four ways during treatment of Dunvallo, Belinus, and Brennius. For the *Short Metrical Chronicle*, see ch.10, ll. 216–224 in standard version; ll. 229 in Royal MS redaction; see pp. lviii, cxxx for discussion.

[31] For Matthew Paris and the Wendover *Flores Historiarum*, see Paul Binski, *Westminster Abbey and the Plantagenets: Kingship and the Representation of Power* (New Haven, 1995) pp. 121–40, esp. 121–22 and 126–28; see also Binski (1986) pp. 39–40 and Lewis (1987) pp. 12–13, 53–57, 75–77, and 135–37. On Paris's description of himself as one having read historical annals, see Richard Vaughan, *The Illustrated Chronicle of Matthew Paris: Observations of Thirteenth-Century Life* (Cambridge, 1993) p. 59. For the copy of Geoffrey of Monmouth (MS Royal 13.d.v) which still exists "with annotations in Paris's handwriting," see Vaughan (1958) pp. 104, 129, 187; also Lewis (1987) p. 482 n.10. On Matthew Paris, Henry Huntingdon, and Roger Wendover, see e.g., Lewis pp. 13, 485 n. 102, 480 n. 28. Also Antonia Gransden, *Historical Writing in England c.550 to c.1307* (Ithaca, 1974) p. 367; Nancy F. Partner, *Serious Entertainments: The Writing of History in Twelfth-Century England* (Chicago, 1977) pp. 197, 37; see also discussion below. For the observation (much in line with the arguments I have developed independently) that Matthew was "clearly reading a copy of Geoffrey of Monmouth's History of the Kings of Britain when he penned the sketch of the ancient roads," see Delano-Smith and Kane, p. 247; also pp. 16–17.

[32] Geoffrey of Monmouth III.5; on *leode*, see Rosamund Allen, ed., *Lawman's Brut* (London, 1992) p. xxviii. On Roman, Saxon, Pict (et al.) invasions, note incidentally that precisely this lamentation of plagues appears likewise in the short preface to Gervase of Canterbury's *Mappa Mundi* or three-column list of religious houses in Britain; for reference see chapter one.

[33] V. H. Galbraith, "Good and Bad Kings in Medieval English History," *History* 30 (1945): 119–132; John and Edward I constitute "the two cardinal reigns." For Arthur, see discussion and notes below.

[34] Gale's transcription (to which I should surely defer) reads "meridies" for "orient," but see discussion just below of the *Scema*'s inscription, straddling the very line of the vertical road, which details the route of Ykenild Street *ab oriente in occidentem*.

[35] Mary Carruthers, *The Book of Memory: A Study of Memory in Medieval Culture* (Cambridge, 1979); see especially chapter one ("Models for the Memory") and chapter three ("Elementary Memory Design"), and pp. 11, 17, 21, 27–28, 101–102; see also note below.

[36] Geoffrey of Monmouth III.5.

[37] For Psalter preliminary illumination sequences, see discussion and notes to chapter one.

[38] Geoffrey of Monmouth III.5; for discussion see Guest pp. 100–101, who regarding the Monmouthshire road says "I cannot find the slightest corroborative evidence to show that this was one of the four great Roads, and I believe it was selected by this unscrupulous writer, merely because it was an ancient highway, and passed through his native country."

[39] See Tatlock p. 12; also p. 281; Henry of Huntingdon text is from the *Historia Anglorum* Rolls Series edition, ed. T. Arnold (London, 1879) p. 12; on the "written laws of the land" to which Henry refers—the *Leges Willelme* and the *Leges Edwardi Confessoris*—see discussion below.

[40] See Carruthers, esp. pp. 43, 116, 109 for her defining metaphor of the well-trained memory as "a powerful portable library" or "a book containing both text and

glosses." Indeed, better to rely upon memory, Carruthers notes, than upon "the cumbersome, inefficient and chancey method of turning the pages of a book," p. 85. Also note Carruthers's point that "the proof of a good memory lies not in simple retention but manipulation"; Matthew's tabulated memory shows itself useful in this sketch not in its ability to "reproduce something exactly" (this would have meant no cartographic sketch at all) but rather in its ability to manipulate data as needed—"to move it about instantly, directly and securely"; pp. 21, 19.

41 On "rude" and critically dismissed, see notes above; also discussion and notes below on the *Scema*'s "error" in bringing the Four Ways together at Dunstable. For the status of virtually any medieval cartographic endeavor as a "bold conceptual initiative," see notes to chapter one; see above for "cartographic revolution." For well-received consideration of the relationship between maps and written texts—with attention especially to questions of composition/construction—see the various essays of Patrick Gautier Dalché (work with which I am only just becoming familiar, unfortunately).

42 For symmetry, see Lewis (1987) p. 323; for the frequent and cross-cultural association of the circle (and to some extent the oval) with things sacred, see for example Rudolf Arnheim, *The Power of the Centre: A Study of Composition in the Visual Arts* (Berkeley, 1982); also Rosalie Vermette, "*Terrae Incantatae*: The Symbolic Geography of Twelfth-Century Arthurian Romance," p. 154, in the context of Arthur's Round Table, on which see below, pp. 153–54. Giraldus Cambriensis's 5" x 4" map of Western Europe (in his *Topographia Hibernia*) is sometimes cited as an early example (c.1200) of mapping at a regional scale. However, the tenor of Gerald's Map of Western Europe differs completely from Matthew's *Scema* in that, as Crone notes, Gerald's is "a portion of a medieval world map, or is based on such a map" and moreover "appears to be intended as a rough guide for the traveller from the west to Rome." The route which forms the map's itinerary begins at Winchester; the map is oriented east. See Crone (1961) p. 14.

43 On Gough's physical characteristics including rusty nail holes and places where names have apparently been wiped away by the sleeves of its users, see Parsons pp. 2–3; for Hereford's nail holes, see P. D. A. Harvey (1996) pp. 11–12; p. 10 for its compass grounding at center. Harvey has called for a more extensive physical examination of this map.

44 On Gough as "an official map" perhaps "kept at London" see Parsons pp. 2, 15; for the suggestion of its being "kept at Westminster . . . for reference purposes" see R. A. Pelham, "Early Maps of Great Britain II: The Gough Map," *The Geographical Journal* 81 (1933), pp. 34–39, esp. p. 39. More recently B. P. Hindle, "Roads and Tracks," in *The English Medieval Landscape*, ed. Leonard Cantor (Philadelphia, 1982) pp. 193–217, has supported this consensus view that Gough "appears to have been an official compilation for government use," p. 196. Hindle adds, however, the intriguing detail that one Robert of Nottingham, while buying wheat for the king in the vicinity of the River Trent in 1325, "appears to have had an early edition of the Gough Map with him, as he gives the same mileages [in his written itinerary] as those on the map," p. 198. On Gough as a model for later maps see for example the incontrovertible evidence of the Beinecke Map (references given above); on the continued relevance of a medieval road map, insofar as the medieval road system itself long continued to be sufficient, see Hindle pp. 214, 195. For recent treatment of Gough, including mention of most of the issues treated below—

i.e., date, provenance, function, manner of production, "national scale," route network, possible lost versions, long-running preeminence, and overall status as "an enigma"—see Delano-Smith and Kane, pp. 47–48, 19–20.

[45] See Crone (1961) p. 17; Jervis pp. 83–84; P. D. A. Harvey (1987) p. 496; Parsons p. 5. On roads as Gough's definitive feature see especially Crone and Harvey; on the relationship of Gough's road network to the island's shape, see esp. Parsons pp. 7, 15; also Hindle (1988) pp. 12–13 and Helen Jewell, *The North-South Divide: The Origins of Northern Consciousness in England* (Manchester, 1994) p. 107. For a diagram sketching Gough's route network onto modern outline-map of Britain, see Delano-Smith and Kane p. 47. On a debt to portolan charts, see P. D. A. Harvey (1996) p. 38. Hindle (1982) has established that precisely those "certain areas of the country [which] were rarely visited by monarchs" are those which are road-poor on Gough; pp. 193 and ff., esp. 199.

[46] On Gough and lines of travel, see esp. Frank Stenton, "The Roads of the Gough Map" (in Parsons pp. 16–20), and Hindle, "The Towns and Roads of the Gough Map," *The Manchester Geographer* 1 (1980) pp. 35–49, but also Crone (1961), Pelham, Parsons, and Harvey (1987), esp. p. 496. For Richard Gough's early assessment, see for example Crone p. 17. The confidence which Parsons displays in Gough as self-evidently "a map for travellers" (p. 15) is picked up with enthusiasm by Hindle a quarter century later, for whom this view is implicit and very much assumed; see for example Hindle (1982) p.196; also Hindle (1980).

[47] On dismissive treatment of the *Scema* overall, see discussion and notes above. On the Four Ways "quite wrongly intersecting at Dunstable," see for example Vaughan (1958) p. 244 and Hindle (1988) p. 13, who notes another "half a dozen fundamental errors." These attitudes may be on the way out, if Delano-Smith and Kane's treatment (2001) is any indication.

[48] See in general Ivan Margary, *Roman Roads in Britain* (London, 1955), Thomas Codrington, *Roman Roads in Britain* (London, 1903), and Guest. For Dunstable see the immediately persuasive evidence of Margary's Maps 5 and 7b, pp. 128 and 193; also 179 (for the connection to Icknield Way); see also Codrington, pp. 62 and 199 for Dunstable, "where Icknield Way crosses Watling Street"; Guest p. 106 notes that the earliest mention of the Icknield Way comes in connection with an account of the founding of Dunstable Priory; see also p. 114. Also on the point of Dunstable as the center of the system, note Harley and Woodward, who observe that "the centering of maps on a particular point of sacred or secular importance" constitutes a paradigmatic if unsurprising "manipulation of the geography of the map to fit a specific perception of the world," p. 505. Geoffrey of Monmouth, as we have seen, provides precedent for the practice of rerouting famous highways through a historian's own locality; see Guest p. 101 and Codrington p. 21 for more.

[49] Stenton p. 16; Hindle (1982) echoes Stenton almost exactly on this point; see also "The Road Network of Medieval England and Wales," *Journal of Historical Geography*, vol. 2 (1976) pp. 207–221 and Hindle (1980). In specific Hindle concurs with the judgment that the road system "made possible a centralization of national government"; roads do "reflect the centralization of government, and there is clearly a national road system radiating from London"; pp. 194, 196–7.

[50] See Parsons pp. 4–5; see also the oval view of London on Paris's Itinerary to Apulia, fig. 2.5.

[51] See Margary p. 27 for regional hubs; p. 53 for London as principle center. For the continued serviceability of Roman roads during medieval times, see A. L. Poole, *From Domesday Book to Magna Carta 1087–1216* (Oxford, 1953) p. 78, who notes that "the roman roads had indeed been so well constructed that even in spite of neglect they were tolerably serviceable" into the thirteenth century, although "their utilization was only piecemeal, given changed settlement patterns." See also Hindle (1982) p. 193, who adds that "a large part of [the roads in the Roman system] were still in use." For how roads "form the most useful and lasting monument to the Roman occupation here," see Margary p. 27; similarly Codrington p. 11. For an ingenious layered reconstruction of known and posited medieval routes, plus other diagrams, see Hindle (1982) pp. 203, 205, 206.

[52] The textual problems surrounding all of this are considerable. In general see Kantorowicz pp. 345 and ff. as well as H. G. Richardson, "The English Coronation Oath," *Speculum* XXIV (1949), pp. 44–75, esp. 61 and ff. On Edward the Confessor, nb. M. T. Clanchy, *From Memory to Written Record: England, 1066–1307* (Oxford, 1993) p. 318: "The conclusion to draw is that most documents extant in his name are of doubtful authenticity." See Tatlock p. 278 for "one to conjure with"; for the Norman practice of confirming the laws of previous rulers (on which see also Poole p. 386); and for the appearance of Fosse et al. in Anglo-Saxon charters. For an illustration, see *Annals of Roger de Hoveden*, esp. pp. 538–540.

[53] Hoveden, pp. 545–47 for reference to the "four Royal public Roads," and for the distinction of "lesser roads," etc. For the extension of the "idea of the 'king's highway'" see Poole pp. 78–80; also Guest pp. 99–100, who notes that "thus the question which were the lines of traffic that once exclusively enjoyed this protection, became one of merely antiquarian interest."

[54] Many have noted the long-standing link between roads and maps; see for example Jewell p. 106 ("the road system was a prime interest of early mapmakers"); also Margary p. 21 who notes that extant Roman itineraries "follow in general the principal roads"; also Crone, *Maps and Their Makers* (London, 1953) p. 5, who has speculated that ideally, perhaps, a Roman map was primarily "a map based upon the foundations of the road system." On the question of royal highways "fanning out from a metropolitan hub" as *themselves* an emblem for the state, I find a comparison introduced by Tatlock too rich to resist including: the Four Ways' symmetrical arrangement into vertical, horizontal, and two diagonal lines, she notes, is "like a Union Jack," p. 281. In this spirit of contemporary comparison, the resemblance between Britain's Roman/medieval road system and its modern railway system might also be noted, on which note see de Certeau's seminal discussion of railways in *The Practice of Everyday Life*, pp. 111 and ff. Like the railway after it, the medieval road system both "reflects" and "makes possible" a "centralization of national government," as Hindle (1982) has said, pp. 194, 196–97.

[55] See previous note on the link between maps and roads. Parsons, incidentally, has suggested that "it is not beyond the bounds of possibility that the compiler of the Gough map was acquainted with a road map or itinerary in the later Roman style, or at least that contemporary conventions were based on such a source," p. 10.

[56] On breakdown into discontinuous sections, see Margary pp. 16, 22–24, also Poole p. 78; Margary p. 25 for main radial center; p. 20 for regional hubs; Codrington p. 11 for different surveyors.

[57] On straightness see Margary p. 12; Codrington p. 29 suggests that their straightness may be "too much insisted upon" but this point is certainly much insisted upon by medieval writers—on Belinus's straight roads, for example, see Geoffrey of Monmouth III.5/II.18. For the *agger*, ditches, and construction overall, see Margary pp. 18–22, who notes that while usually the *agger* is "quite a modest ridge," in some cases "the amount of apparently unnecessary labour put into these embankments at some points must have been prodigious, and it is difficult to believe that it would have been done by such practical engineers without good reason. In some cases, as on Ackling Dyke in Dorset, the road was passing through a district much occupied by native Britons, and it seems possible that the construction may have been done to inspire them with the might of the conqueror, but this hardly seems an adequate explanation generally." See also Codrington pp. 13–16 on construction, who stresses that the Romans made roads "according to the situation, and materials available," and that they were only sometimes paved; often their eventual destruction, he adds, was for the sake of their materials.

On the placement of roads, see Margary pp. 20–21, who notes that Roman roads are typically "remote from present village sites, which lie a mile or two off the routes, to a remarkable degree." Codrington p. 31 makes same point; see also Alfred Watkins, *Early British Trackways* (Hereford, 1922). This removal from village sites would have only reinforced the sense of Roman roads as a technology of domination rather than of local utility.

[58] As Margary notes, "the layout of the roads was generally undertaken at first in response to military requirements for the conduct of a campaign of conquest or for the effective policing of a country after its subjection had been accomplished," p. 12.

[59] On the *cursus publicus*, the transportation system set up by Augustus, see sections by O. A. W. Dilke in Harley and Woodward (1987), esp. pp. 239, 236; see also notes and discussion (regarding the Peutinger Table) above. On the *Corpus agrimensorium*, see Egerton p. 21

[60] See Egerton esp. p. 15; for similar observations by Margary and Codrington see notes above; nb. also Egerton's proposal that "every human society anywhere in the world . . . has believed that geometric patterns formed in orthogonal relationships not only pleased the eye aesthetically but possessed talismanic power," p. 11.

[61] See notes to chapter two, above.

[62] See Davies (1990), p. 113.

[63] Elizabeth Salter, *England and International: Studies in the Literature, Art, and Patronage of Medieval England* (Cambridge, 1988). On "patriotism" as a "factor in politics" following 1258, see M. T. Clanchy, *England and Its Rulers 1066–1272* (Oxford, 1983) pp. 14–16 and 262; on fiscal ratification/taxation, see p. 259 as well as reference in note to chapter one, above. For the Treaty of Paris, pp. 223–224, on which see also Maurice Powicke, *The Thirteenth Century 1216–1307* (Oxford, 1953) pp. 84, 125–28.

[64] See especially Davies (1990) p. 121; for the general "absence of political masterminding" in medieval campaigns of conquest, see also Robert Bartlett, *The Making of Europe: Conquest, Colonization and Cultural Change* (Princeton, NJ, 1993) p. 262, who

allows, however, that Edward I may constitute something of an exception to this; see discussion and note below; also Davies (1990) p. 124.

[65] Davies (1990) pp. 123, 126; Powicke p. 129.

[66] Kantorowicz cites the cases of Philip III's immediate accession upon the death of Louis IX in Africa in 1276 and of Edward I's smooth and immediate assumption of full regal power in 1272, despite a delay in the actual ceremony of his coronation until 1274, as evidence suggesting that by this time the "live essence of liturgical kingship had evaporated." In short, although "the coronation ceremonial . . . was not abandoned" and still served "as a medium for the quasi-religious enhancement of the dynasty," the central action of the king's anointment no longer held the importance it did, and thus "the ancient idea of liturgical kingship gradually dissolved"; pp. 318, 329–330, 192.

[67] Overall see Poole, esp. pp. 476–83; see also Elizabeth Hallam, ed., *The Plantagenet Encyclopedia* (London, 1996) pp. 113, 125, who notes that some hundred years later cries of "Louis of France!" were heard in the streets of London, at the most tangled moments of Edward II's reign (1326–27).

[68] See Stephen J. Herber, Jr., "The Vercelli Book: A New Hypothesis," *Speculum* X (1935) pp. 91–94 for a summary of the arguments for, and then an argument against, Guala's involvement with the transport of this manuscript to Italy. For the Vercelli Map, see discussion above, esp. P. D. A. Harvey (1996) pp. 30–33, 27 and Ernst Kitzinger, "World Map and Fortune's Wheel: A Medieval Mosaic Floor in Turin," *Proceedings of the American Philosophical Society* 117. 5 (1973) pp. 344–373, esp. pp. 358–61 and note 89. The argument might be made that the heavily apostolic nature of the Vercelli Book's selections would fit well with the theme of such a *mappamundi* as Vercelli, which like the cartographic floor mosaic at Turin has important links to Beatus manuscripts. Guala also, incidentally, spent significant time in the west of England (for example, during the period of Henry's coronation at Gloucester)—that is, in the region where the revival of interest in Anglo-Saxon characterized by Layamon and the Tremulous Hand of Worcester had its base. On Henry's struggle as a Holy War, see Clanchy (1983) p. 203; on Henry's coronation at Gloucester, see Powicke pp. 1–3.

[69] See overall Kantorowicz, esp. pp. 347–348, and Richardson pp. 46–48; nb. especially Richardson's insistence that "no one regarded the coronation oath as fixed and invariable"; that "no doubt" Edward I "took an oath going beyond the traditional three promises and engaging him to maintain the rights of the crown unimpaired"; and that there are "similar references" to a like promise made by Henry III in 1233 and 1235 though not by Richard I or John; see pp. 51–53. For Edward I and his references to the "additional promise" he took at his coronation in nine documents between 1275 and 1307, including one regarding his asserted right to perambulation of forests and another regarding the Scottish conflict (both in 1301), see Richardson pp. 49 and ff.

[70] See Kantorowicz p. 353.

[71] See Richardson pp. 56, 74; see also Powicke p. 2 on John's symbolic surrender of the realm to the papacy.

[72] Overall see Richardson pp. 54 and esp. 60–62. On St. Edward's Laws, "doubtless there had been much loose talk" as well as "documents going about" so that "to the minds of some men, at least, this was a reference to a perfectly definite book quite commonly met with."

[73] See Davies (1990) p. 124: "when in 1254 the Lord Edward was given an endowment which included all the royal territories in Wales and the lordship of Ireland it was specifically decreed that these lands were 'never [to] be separated from the crown, but should remain entirely to the king and England forever.'"

[74] See Kantorowicz pp. 345–46; also R. William Leckie, *The Passage of Dominion: Geoffrey of Monmouth and the Periodization of Insular History in the Twelfth Century* (Toronto, 1981), who notes that "there can be little doubt that for Geoffrey the crucial importance of united rule was an inescapable lesson of both British and Anglo-Norman history," p. 57. For a prime example of the problems inherent in divided rule, see Geoffrey's account of the brothers Brennius and Belinus (Book III). For Gerald of Wales's comparable reference to Henry II's having "included the whole island of Britain in one monarchy, even as it is enclosed by the sea," see Davies p. 78.

[75] On matters Arthurian, see discussion below.

[76] On the *Leges Anglorum*, see Kantorowicz pp. 345–46. For Magna Carta's eventual replacement of the Laws of St. Edward, see Poole p. 477; he cites Maitland's judgment that it is not until "the last years of the twelfth century" that "law becomes articulate," p. 385. See also pp. 5–6 for Magna Carta and its implications generally.

[77] Overall see Binski (1995) esp. pp. 121–122 and ff.; for brief overviews of Matthew and Roger and their relationship, see *The Plantagenet Encyclopedia* pp. 150, 177. The Westminster *Flores* is "never fully representative of an official political history of the centre," Binski has said, but it displays "every sign" of a disposition "to promote royal causes and to present a justification of the reigns of Henry III and Edward I."

[78] Binski (1995) pp. 126–128; the full designation of the document is Chetham Library MS 6712, afterward B.L. Loan MS 94. The compilatory works of both Peter Comestor and Honorious of Autun have links to medieval English *mappaemundi*, incidentally (the Duchy of Cornwall Map and the Sawley Map, respectively). Regarding the *Flores Historiarum*, Leckie p. 98 notes that Roger of Wendover in particular among English chroniclers managed to blend the genres of universal and insular history, and gave equal status to Geoffrey of Monmouth as an authority, such that "the regular insertion of Galfridian sections does not seriously distort the familiar contours of universal history"—although it does reorient matters significantly. On the Chetham St. Edward's close correlation to the Painted Chamber mural, Binski notes that while discrepancies do exist these merely "point to the essentially fluid character of English royal images," p. 128.

[79] See Binski (1995) p. 126.

[80] See Powicke pp. 437–44.

[81] As Tout reports in his entry on Dafydd ap Gruffudd in the *Dictionary of National Biography*, "an unseemly contention between the representatives of York and Winchester for the right shoulder resulted in the triumph of the southern city. The other quarters were exposed at York, Bristol (or Chester), and Northampton." For Llewelyn ap Gruffudd, William Wallace, and Simon Fraser see Powicke pp. 428–29, 712, and 716 respectively. For further discussion of the links between the medieval highways and the practice of drawing and quartering, in the context of an incipient public sphere, see Wendy Scase, "'Strange and Wonderful Bills': Bill-casting and Political Discourse in Late Medieval England," *New Medieval Literatures* 2 (1999) pp. 225–248.

[82] See overall R. S. Loomis, "Edward I, Arthurian Enthusiast," *Speculum* 28 (1953) pp. 114–127. On coronation, see discussion and notes above. Regarding Round Table celebrations, Geoffrey Ashe, *The New Arthurian Encyclopedia*, ed. N. J. Lacy (New York, 1996) pp. 518–19, notes that Edward I attended at least five such functions and "organized a lavish one himself in 1295 to celebrate his second marriage"—an event which may have occasioned the initial construction of the (still extant) Winchester Round Table, on which see discussion below. "Quite a number of Round Tables were held during the Middle Ages," Ashe notes, describing them as "a type of aristocratic festival" in which "nobles played Arthurian roles, danced, banqueted and competed in jousts." On the Round Table held in 1284 at Nefyn, see Loomis p. 117; also Powicke pp. 429, 515; Salter pp. 93–97; R. R. Davies, "Edward I and Wales" in Trevor Herbert and Gareth Elwyn Jones, eds., *Edward I and Wales* (Cardiff, 1988) pp. 1–10, esp. pp. 5–6; and N. Denholm-Young, "The Tournament in the Thirteenth Century," in *Studies in Medieval History Presented to F. M. Powicke*, eds. R. W. Hunt, W. A. Pantin, and R. W. Southern (Oxford, 1948) pp. 265–66. The Round Table held at Nefyn was an event noted by contemporary chronicles including the *Flores Historiarum*. The unmistakable association between the 1284 tournament at Nefyn and *royal* political maneuvering, it is interesting to note, marks a complete reversal or reorientation of the function of the aristocratic tournament as a political institution at the start of Henry III's reign, when during the push toward Magna Carta the tactic of calling frequent tournaments had served as a means for keeping baronial forces ready in arms; see discussion above.

On the Crown of Arthur, the Stone of Scone, and the Shrine of St. Edward, see Loomis p. 117, Powicke, 515–16 and Davies (1990) pp. 125–26 as well as Binski (1986) p. 111 and Binski (1995), pp. 105, 135, 138. No hard evidence survives to suggest whether Edward I in fact actually re-enacted his coronation ceremony as planned, with newly won crown and himself in the role of Arthur. But Edward clearly "liked to think of himself as *Arthurus redivinus*," as Loomis p. 126 has said, and overall seems to have pursued an identification with his legendary predecessor something in the manner of Henry III's attempts to root his kingship bodily in the person of St. Edward the Confessor. Interestingly on this point, the Douce version of the *Flores Historiarum* includes a full-page illustration which appears clearly to adapt the Painted Chamber and Chetham *Flores* scene of St. Edward's coronation—but replacing St. Edward with King Arthur. See Binski (1995) p. 126; also fig. 166 on p. 121.

[83] See Loomis esp. pp. 114–15, who notes that "a cult of Arthur" seems to have flourished in "most of the aristocracies of Christendom." Yet in England, "it is only with Edward that we possess a series of indications . . . of a strong concern with both the historic and romantic traditions of Arthur." See also Tatlock p. 436 on Geoffrey's *Historia*. Incidentally, Arthurian romancier Rusticello of Pisa—co-author of Marco Polo's *Travels* (c. 1299?)—enjoyed Edward's patronage in the early 1270s, and may have accompanied him to Palestine. Working from a volume of French romances which Edward had brought with him to the East—significantly, the only book known for certain to have been owned by Edward, with the exception of a manual on warfare—Rusticello compiled the first Arthurian romances by an Italian in 1270–71. See *The Travels of Marco Polo*, ed. Ronald Latham (London, 1958) esp. p. 4. That Prince Edward should have brought such a book with him on his passage to the East suggests not only, as Salter (pp.

93–95) has observed, something about the willingness of Latin kingdoms in the East in this period to adopt Arthuriana in their rituals of court. It also suggests the presence, from an early stage, of important Arthurian influences in Edward's conception of the kingship he would assume (and begin to transform) just a few years later.

Where Powicke reads Edward's Arthurian enthusiasm as "dictated by expediency, by political considerations," Loomis insists that "sentiments too are involved," p. 114. But see especially Salter pp. 93–95 for discussion of Edward's "calculated Arthurian display." Salter cites Rosalind Wadsworth on the "absence of Arthurian subjects in the Anglo-Norman romances written under the patronage of the Barons" c. 1170–1300, suggesting that this "reflects the consciousness of a 'royal and Arthurian' cult: the deliberate avoidance of that area of narrative by writers catering for a class of patrons who were frequently in opposition to the crown seems entirely likely." It may be useful to note, in the context of the above, that in twelfth-century France, on the other hand, Arthurian literature tends to connect not with royal but with regional aristocratic causes and interests, as evidenced by its production at provincial courts like Troyes. See generally Erich Köhler, *L'Aventure Chevaleresque: Idéal et Réalité dans le Roman Courtois* (Paris, 1974), pp. 7–43, and Ad Putter, *Sir Gawain and the Green Knight and French Arthurian Romance* (Oxford, 1995).

Finally on this question, see the reproduction (their fig. 2.14) and fascinating mention in Delano-Smith and Kane of a fifteenth-century map of "'the whole of England' (*Totius Britanniae tabula chorographica*)" which not only possesses various features suggesting "an affinity with the Gough Map" but—intriguing to relate—appears in a manuscript of none other than Geoffrey of Monmouth's *Historia Regum Britanniae*: London, B.L. MS Harley 1808, f. 9v, c.1400. This manuscript also includes "a symbolical representation [or plan] of the city of York," an illustration probably "intended as a commemoration of the role York played in the conversion of Britain"; p. 22. I have not been able to incorporate this information into my arguments explicitly, but the implications of such a map in such a place should be generally apparent.

[84] On Glastonbury, see Loomis p. 116 who speculates as to the "glamourous associations" likely present in "those white bones" for a "reader of romances" such as Edward; he sees "mere disinterested curiosity" as much in play as political calculation. On Edward II and Geoffrey's *Historia*, see pp. 122–23. 1301 was the same year in which royal clerks borrowed heavily from the *Historia* in support of Edward I's claim for legal overlordship in Scotland, in a famous letter to Pope Boniface; see discussion below, chapter three. Edward II was knighted five years later, during another Arthurian festival, at Westminster, not long after news had arrived of a new uprising of the Scots (swearing, per romance convention, not to sleep two nights in the same place until the enemy had been conquered). Young Edward had been born in 1284 during the king's long residence in Wales (hence *of Caernarvon*), a few months before the conquest had been marked with the Round Table at Nefyn.

[85] See for example M. R. Shichtman and J. P. Carley's introduction to *Culture and the King: The Social Implications of the Arthurian Legend* (Albany, 1994) pp. 4–14; this collection aims "to distance [itself] from the notion of a consistent and recoverable Arthurian past. Our aim, instead, is to study the legend as a set of unstable signs appropriated by different cultural groups to advance different ideological agendas"; different

formulations of the material, that is, have "the potential to promote the imperatives of widely divergent social groups," pp. 4–5.

[86] Vermette, pp. 149–51; consider also the perhaps obvious but foundational reminder that "the physical space of the Arthurian cosmos is centered on Britain," p. 147. On the project of bringing order to the outlying bounds of Christian civilization, recall the similar conviction of thirteenth-century crusade-advocates including Robert Grosseteste and his protégé Roger Bacon that successful Western reconquest of Jerusalem depended directly on a prior establishment of order and peace *within* Christendom, in which project geographical learning played a crucial part. Bacon is one of that string of Western scholars who sent a world map to the pope in the hopes of moving him to a more vigorous exhortation of crusade; this map, apparently a mathematically oriented one, is not extant. See note to chapter one, above, as well as Egerton pp. 12, 29–30 and Woodward (1989) p. 8 on Grosseteste and Bacon.

[87] See Vermette, p. 154; further, "the king's restoration to health will then be mirrored in the restored health and prosperity of the land and the court." This link between the health of the king's body and that of his kingdom has numerous examples beyond the Arthurian. But that it is basic to a specifically Arthurian conception of kingship is suggested by even the marginal notes to the British Easter Annals, where plague is understood to be the direct result of a historical Arthur's death. As Vermette points out, "next to AD 539 is found 'the strife of Camlaan in which Arthur and Modred perished. And there was plague in England and Ireland'"; pp. 148, 154.

[88] For discussions of dating and a likely connection to Edward I, see Sue Ellen Holbrook, "Malory's Identification of Camelot as Winchester," in *Studies in Malory*, ed. James W. Spisak (Kalamazoo, 1985) pp. 13–27, esp. pp. 20 and 26 note 23; Ashe pp. 518–19; and *Age of Chivalry: Art in Plantagenet England, 1200–1400*, ed. J. Alexander and P. Binski (London, 1987) pp. 311–312. All of these commentators reject the "conjecture . . . that it was made for Edward III" in favor of the late thirteenth century. Regarding Edward III see for example Thomas Beaumont James, *The Palaces of Medieval England, c.1050–1550* (London, 1990) pp. 119–120 and fig. 69. There is no reason why an already existing table cannot have been connected very directly with Edward III's establishment of his Order of the Round Table in 1344.

E. W. Tristam, in his *English Medieval Wall Painting* (Oxford, 1950) pp. 180 and 610, suggests that such a table may date ultimately to 1236. Tristam's grounds for this are references to a painted Wheel of Fortune commissioned for Winchester Hall in the *Liberate Rolls of Henry III 1226–40*, p. 405, as well as his formal observation that the Table's "radiating divisions" and the king at its head "recall the general composition of a Wheel of Fortune." But even were Winchester's Round Table originally a Wheel of Fortune, being later transformed in theme and function, carbon dating mitigates against this suggestion of an early to mid-thirteenth-century origin. Nonetheless, the supporting evidence Tristam marshals bears repeating. Bishop of Winchester Peter of Roches, who claimed to have met up with King Arthur while lost hunting in the woods, lent Henry III a substantial sum in 1236 "for the works in the hall at Winchester Castle"—where the Wheel of Fortune in question was painted later that very year. See Tristam p. 17 for the anecdote. See also Kitzinger, overall, on Wheels of Fortune in the context of cartographic imagery, esp. p. 365 n.140 for discussion of Winchester.

On Winchester's Round Table having once had legs, see Holbrook. On the c.1521 repainting, see Tristam p. 180 and esp. T. B. James p. 26, who connects this to a formal visit to Winchester Hall by Holy Roman Emperor Charles V: "On entering the building the visitors saw King Arthur's Round Table hanging on the wall. . . . The frail Tudor dynasty drew strongly on the long established Arthurian legend to bolster its image with the visiting emperor." My transcription of the knights' names follows P. J. C. Field, *Notes and Queries* 25:3 (1978), p. 204. Field and Holbrook have shown that the Table as painted now appears to follow a passage from Sir Thomas Malory in its listing of the twenty-four knights, and that Malory for his part had insisted upon an unlikely identification of Winchester with Camelot, apparently due to his knowledge of this unique artifact there. The year after publication of Caxton's Malory, incidentally, Henry VII's son—named Arthur—was born at Winchester; see Holbrook p. 27, note 29. The idea was that he would rule as Arthur II; he died early but had he lived his placement where Henry VIII ends up, at the center with the legend King Arthur directly above his head, would have been even more appropriate and effective as propaganda.

Finally, for treatment fully extensive and now surely authoritative, see Martin Biddle, ed., *King Arthur's Round Table: An Archaeological Investigation* (2000); unfortunately I encountered this study too late to incorporate it here.

[89] See for example the comments of David Harvey, *Explanation in Geography* (London, 1969), pp. 373–76 and ff.

NOTES TO CHAPTER THREE

[1] On mapping and territorial expansion, "whether the context [is] colonial, military, or political," see Harley and Woodward p. 507; also Schulz p. 447. On the tendency of military and administrative ambition to outrun practice, see for example Davies p. 123.

[2] On the "fisc" and "a monarchy of the whole island," as well as on the (presumed) opposition between "factual geographic" and "symbolic," see discussion in chapter two, above.

[3] See chapter two ("The Ideology of a Road Map") for Gough's roads overall and for roads and maps in general. By way of contrast to what the Gough Map provides, Hindle (1982) points out that "one of the five most travelled routes" in the itinerary of Edward I is "from Northallerton to Newcastle, which reflects the seven Scottish campaigns," p. 215; the other four of the top five went out from London.

[4] On routes omitted from Gough but "known from other sources to have been in use," see Stenton pp. 16 and ff.; on the *Canterbury Tales*'s lack of reference "either to the road or to the state of the road"—evidence which "might be taken to imply that the road was well known and in good condition"—see Hindle (1982) p. 196. See also discussion just below on absence.

[5] On obscure towns such as *Bilefeld* between Stamford and Lincoln, see Stenton p. 16, who generalizes that "many important secondary routes are only indicated upon the map by the names of what Leland would have called 'thoroughfares,' that is, villages or small towns through which a long distance road runs." This is, incidentally, exactly the practice of the Hereford Map, at least according to the argument of Crone (1954/1965). In addition to London-Dover, other important "thoroughfares" on the Gough Map (i.e.,

routes lacking a line) include London to Ipswich and York to Newcastle. This occasional absence of lines to mark a road may ultimately be put down to the notion that, arguably, the medieval "concept of a road" meant "more a right of way than a physical track"; see Hindle (1982) p. 208.

[6] For detailed and sometimes exasperated discussion see Crone (1961) p. 17, Stenton pp. 16–20, and Pelham p. 38, all of whom speculate regarding modern correspondences for the Gough Map's given distances. Their shared irritation derives from the fact that the scale of measure seems not to be constant, but rather fluctuates unreliably; each writer does resist the temptation to denounce the mileage figures "as merely [an] estimate" or "the guesswork of a man to whom the roads of England were only known by vague repute"; Stenton pp. 19–20 allows for "the general medieval tendency towards computation in round numbers" and notes that 40% of Gough's distances are given as five, ten, fifteen, or twenty miles.

[7] For Edward I's 1295 expedition into Wales, see Powicke pp. 441–443, who describes it as a particularly "rapid and massive demonstration of royal power"; it included another occupation of Anglesey and construction of a new castle at Beaumaris, and in the course of the king's military circuit of Wales "all resistance crumbled away and the Welsh returned to his peace. By August he was back in Westminster." For details on Gough's line, the route along which Edward I passed but which "differs widely" from the route as known from the sixteenth century onward as well as from its Roman predecessor (on which see Margary pp. 13 and ff.), see Stenton p. 17; see also J. E. Morris, *The Welsh Wars of Edward I* (Oxford, 1901) pp. 264–65. See Hindle (1982) p. 208 for Gough's roads' correlation with the routes of Roman highways.

[8] For discussion of Gough's hill and mountain icons, see Parsons p. 8, who suggests that the relative lack of attention to highlands, compared for example with rivers, towns, and roads, is "perhaps because they hindered rather than assisted communications or perhaps because they would encumber the map and make it more difficult to read." See discussion below on this question of wayfaring versus other uses of the Gough Map, however. Parsons finds *Snowdonne*'s cogged wheel "a degenerate form of the more realistic cone-shaped mountains of the Hereford Map."

[9] See Davies for the castle as both symbol and tool of domination, p. 40; similarly Bartlett pp. 65–84. On Edward's castle building in North Wales see for example Michael Prestwich, *War, Politics and Finance Under Edward I* (London, 1972) p. 29; Powicke pp. 430 and ff.; and Ifor Rowlands, "The Edwardian Conquest and its Military Consolidation," in *Edward I and Wales*, ed. Trevor Herbert and Gareth Elwyn Jones (Cardiff, 1988) pp. 41–72, esp. pp. 44–53. Regarding links between the Gough Map's sites and English-Welsh conflict, Hawarden in Flintshire (for example) is where the 1281 revolt begins; Flint and Rhuddlan are the sites of new castles following 1277, and St. Asaph's is the site of the cathedral whose move to Rhuddlan is proposed in 1281; Carmarthen becomes an administrative center of west Wales post-conquest. Further south, Gough site Llandeilo is the site of an ambush, Llandovery is a revolt site in 1287, Ruthin is the new seat of the Earl of Lincoln and later a site for rebellion in 1294; and so on; see Powicke pp. 419, 412–3, 434, 430, 419, 439, 431, 441. One particularly evocative example is provided by Bardsey, an island filled with text off the west coast of Snowdonia. The inscription here (*bardsey ubi sunt vacinatores*) has been used to help date

the prototype version of Gough to before 1284—as that is when Edward I drove out the island's famous bards. See Parsons pp. 2, 27 and Pelham p. 35; also, note overall Pelham's well-founded "suspicion that the map was not all drawn up at the same time," p. 36.

[10] For Gough's legend on Anglesey, see Parsons pp. 6–7. For the island's fertility, see Gerald of Wales, *The Journey through Wales* and *The Description of Wales* ed. Lewis Thorpe (London, 1978) pp. 187, 230; closely comparable passages appear in each. On English occupation, see for example Powicke pp. 412, 422–23, 426–27, 430 and ff., 443. On the opportunity afforded by Edward's visits for a measurement of the island, see Powicke pp. 6–7; see Stenton p. 19 for Gough's 150 measured road-segments.

[11] For Edward's Chester to Cardigan road, some of it a re-establishment of Roman sections, see Hindle (1982) p. 195; for forest cuts see also Prestwich (1972) pp. 28, 110: "Edward's success depended on the maintenance of good communications, and much time and money was spent in cutting roads through the dense forests of Wales. In 1277 household knights were used to supervise the workmen, and in August as many as 1,800 Axemen were cutting a road to Rhuddlan"; see also Morris, pp. 139, 212, 223–25. For "the deliberately imperial majesty of Caernarvon Castle, with its evocation of the walls of Constantinople and its statue of the king above the great twin-towered gatehouse," see Davies p. 127; also Prestwich (1972), who calls this "the chief material symbol of the conquest of Wales" and notes its links with the Roman past; pp. 170–71, 242. For the road's full route see Parsons p. 36.

[12] On Edward I's 1284 and 1295 perambulations, see Morris pp. 199–203; p. 202 for "populations . . . English" and for the establishment of administration; also Davies p. 123. On the Round Table at Nefyn, see notes to chapter two; note also Powicke p. 516. Hindle (1982) provides several useful diagrams showing the movements of Edward I.

[13] See Hindle (1982) pp. 193, 195; also Hindle (1976) p. 211.

[14] This is computed to have been about 1,618 yards; see the *Oxford English Dictionary* s.v. "mile"; see notes above for references to the slippery question of the Gough Map's distance-measurements and their possible equivalents. For my purposes, the "actual" accuracy and consistency of Gough's measurements is relatively unimportant.

[15] See the *Oxford English Dictionary*, "mile"; also *The Dictionary of Medieval Latin from British Sources*, ed. D. R. Howlett et al. (Oxford, 1989), for the Roman mile, "whose precise length varies by region and time"; see p. 1108 for *miles* and p. 1109 for *mille;* nb. also Codrington's doubtfulness that the so-called "milliaries," or markers along Roman highways which bear inscriptions to emperors, can properly be called milestones, p. 16.

[16] On absence proving nothing, see R. Ian Jack, *Medieval Wales* (Ithaca, 1972) p. 216. For a reading of absence, see David Wallace, "Chaucer and the Absent City" in *Chaucer's England: Literature in Historical Context*, ed. Barbara Hanawalt (Minneapolis, 1992) pp. 59–90. On a "rudimentary survey" to which Gough's prototype was "in all probability closely connected" see for example Parsons p. 15.

[17] For text of the commission cited, see *Calendar of Patent Rolls, 1272–1281*, pp. 342–43; for notices see especially Hilton p. 91, Powicke pp. 358–59, Clanchy (1993) p. 6, and Trevor John's introduction to *The Warwickshire Hundred Rolls of 1279–80* (Oxford, 1992) pp. 2–6.

[18] For text, see references in previous note. See esp. Clanchy (1993) for the surviving returns and potential implications and/or reasons for this, including the proposition that since the extant returns in effect sketch a line between Oxford and Cambridge, this may be put down to the greater availability of clerks in these regions; see also John pp. 2–6. Note also, incidentally, that the Gough Map has a density of settlement-detail in the midlands comprehensive enough to match the Hundred Rolls'.

[19] Wales might easily have been surveyed later, particularly after Edward I established English-style counties with the Statute of Rhuddlan in 1284; but prior to the 1290s there was no opportunity for English survey past Northumbria, nor would there be even in 1306, at the high water mark of English encroachment in Scotland—Edward I held Berwick and other key towns but English authority didn't penetrate much into the countryside beyond, certainly not the highlands.

[20] See discussion and notes above, regarding Gough's provenance, dating, and early use.

[21] Hindle (1982) p. 210; nb. however that almost 40% of Gough's roads run along the lines of old Roman roads, p. 218.

[22] See note above on citation for "go personally," etc.; see Peter Mathias's foreword to John's *Warwickshire Hundred Rolls* p. v for "greatest enquiry."

[23] See especially the Beinecke "Map off Ynglonnd," New Haven, Beinecke Library MS 558, fol. 47v-48r (with accompanying highway chart on f.47r); for discussion see Daniel Birkholz, "The Vernacular Map: Re-Charting English Literary History," *New Medieval Literatures* 6 (2003) pp. 11–77 as well as "The Gough Map Revisited," *Imago Mundi* (in preparation). See also Crone (1961) p. 17, who notes that "it is interesting to know that many of [Gough's measurements] were still in use three hundred years later," for example in Ogilby's *Britannia* of 1675. For the sixteenth century's initiatives in terrestrial mapping, see for example P. D. A. Harvey (1993), E. G. R. Taylor (1930), and Buisseret (1992).

[24] Powicke pp. 358–59; John, esp. pp. 1–6; Powicke pp. 358–59.

[25] *Calendar of Patent Rolls, 1272–81*, p. 343.

[26] On the fisc as a "hall mark of sovereignty" see Kantorowicz p. 189; see also Davies p. 126. I have followed Kantorowicz in employing variations on the term "public sphere" but see J. Habermas, *The Structural Transformation of the Public Sphere: An Inquiry into a Category of Bourgeois Society* (Cambridge, MA, 1989), esp. pp. 4–16, for a comprehensive and nuanced discussion; see also Habermas's "Further Reflections on the Public Sphere" in *Habermas and the Public Sphere*, ed. Craig Calhoun (Cambridge, MA, 1989) pp. 426–61. Overall, Habermas outlines a three-part movement: the emergence under bourgeois humanism of a democratic public sphere based in horizontally organized reading circles, and its subsequent breakdown under the mass-communications assault of twentieth-century capitalism being his major concerns. The first stage in Habermas's model of the public sphere is what he terms "representative publicness," characteristic of the medieval period. Public space in this formulation is used by the Crown as a backdrop against which power may be articulated, through mechanisms such as spectacle and pageantry. This model works well enough for the phenomena under analysis in this study; but other modes of medieval political association and cultural production call Habermas's model of the medieval into question, and begin to refine its monolithic

denial of agency to medieval subjects; see for example Scase (1998), esp. pp. 226–227, 247.

[27] See Davies p. 1 on the tendency to place too much emphasis on battles. For an example of the general use of the Bayeux Tapestry, see Andrew Sanders, *The Oxford Short History of English Literature* (Oxford, 1994) pp. 28–29.

[28] On Welsh town-charters, see Powicke pp. 432–33; also Bartlett pp. 167–182, esp. 173; Davies pp. 12–15; Prestwich (1972) pp. 170–71, 242; and Morris pp. 4–5, 199–203.

[29] See in particular Prestwich (1972) esp. pp. 151, 170; on "essential work" see also Ifor Rowlands, "The Edwardian Conquest and its Military Consolidation," in *Edward I and Wales*, eds. Trevor Herbert and G. E. Jones (Cardiff, 1988) pp. 41–72, who notes that "Welsh independence did not as much perish in a clash of arms as suffocate in a welter of parchment," p. 48.

[30] See A. Z. Freeman, "A Moat Defensive: The Coast Defense Scheme of 1295," *Speculum* 42 (1967) pp. 442–62, esp. 445, 461–2.

[31] See in particular G. P. Cuttino, "King's Clerks and the Community of the Realm," *Speculum* 29 (1954) pp. 395–409, who observes "a wide gradation of functions and importance" among those entitled king's clerks. "Specialization of function has developed to a degree . . . but never to the point where it is unusual for a clerk to be reassigned or taken away from his customary task for a special mission. Taken together, king's clerks constituted a reservoir of carefully trained talent that could be utilized as the situation demanded"; pp. 396–97. On nostalgia for the reign of Edward I and on conquest as orchestrated by the royal household, see also Powicke, esp. p. 340

[32] Cuttino p. 404 has noted that the class of freeholders from which royal clerks were mostly drawn comprised "the backbone of the *communitas regni Anglie*": "There is little wonder that, all told, these clerks did an extremely competent job, for their stake in good government was as great as was the stake of those from whom they came." That is, there was "a community of interest between the civil servant and the class from which he was drawn." Cuttino goes on to speak of the "rarity of feudal factionalism in the ranks of the civil service. Whatever their backgrounds, however close may have been their previous attachments to the households of great magnates, once in the service of the king, by and large they soon fell under the spell of an attraction that Professor Strayer has described as laicization, 'the political aspect of secularization' or 'the development of a society in which primary allegiance is given to lay governments,' and in this instance, to the king." For more see discussion and notes below.

[33] For example, one of the key figures in Edward I's administration was Edmund of Cornwall, the first cousin and seneschal of the king. Son of Richard of Cornwall (King of the Romans from 1257) and following Richard's death Earl of Cornwall himself, Edmund was of course no king's clerk *per se*, rather one of the wealthiest and most influential nobles in the realm. But in the course of a professional life as one of the king's advisors, high officers, and key partisans, Edmund was often closely associated with men like John of Kirkby, a royal clerk who in 1284 headed the next major inquest after Little Domesday (known as "Kirkby's Quest") and who probably played a central role in the earlier and more ambitious survey. As noted earlier, at his foundation of Ashridge College in 1283 Edmund of Cornwall bequeathed (among other items) a *mappamundi*,

now fragmentary, which appears to have been based on Henry III's Westminster world map. Obviously the technical cartographic overlap between this Duchy of Cornwall Map (whose surviving fragment treats Africa) and the Gough Map of Britain is extremely limited. The relevant point here is that it is precisely those high royal officials who had a professional role in surveying the realm who *also* had an interest in cartography—and that these are the same officials involved in administration of Edward's wars of conquest. On Kirkby's Quest see *Feudal Aids, 1284–1431*, vol. I (1899) pp. viii-xxii. On the Duchy Map and Edmund of Cornwall see mentions above as well as Haslam p. 44. As regards the matter of Scotland (discussed at length below), note incidentally that Edward I spent Christmas 1290 at Ashridge, mourning Queen Eleanor and mulling over the Scottish succession crisis in the company, apparently, of the Duchy of Cornwall *mappamundi;* see Haslam as well as Stones and Simpson (1978), for which reference see notes below.

Regarding the Gough Map's early history—prior to its acquisition by Richard Gough in 1774, and subsequent gift to the Bodleian in 1809—see Parsons pp. 1–2, who traces its ownership through Thomas Martin to Peter Le Neve (d.1731): "Le Neve was in touch with government departments and had been appointed one of the deputy chamberlains of the exchequer, an office which he gave up in 1705/06. He would therefore have had the opportunity of obtaining it from an official as well as from a private source."

34 See discussion in chapter two ("Kingship, Part Three: The Fisc"), esp. on "public things" or *res publicae.*

35 As Parsons has reported: "Drawn with relatively widely spaced double lines, [rivers] occupy quite a considerable area of the [Gough Map], and convey the impression that they are a fundamental part of it"; p. 8. For subsequent points see J. F. Edwards and B. P. Hindle, "The Transportation System of Medieval England and Wales," *Journal of Historical Geography* 17.2 (1991) pp. 123–34, esp. pp. 124, 129; see p. 125 for a reading of Gough's rivers in line with Parsons's. See also Poole, pp. 79–81 on the importance of rivers to wayfaring.

36 See the *Annals of Roger de Hoveden*, pp. 545–547 in C.S. Wilkins, *Leges Anglo-Saxonicae* (London, 1721).

37 Parsons calls the Gough Map's attention to riverways "a knowledge which could only have been accumulated over a great many years" but it is unclear as to why this must be, or why those with local knowledge cannot have communicated it to inquiring commissioners just as they did the details of landholding, etc. On Little Domesday's concern with "waters and rivers" see *Calendar of Patent Rolls 1272–1281*, pp. 343–44.

38 See Jewell p. 107 for the beginnings of some dissatisfaction with the received characterization of the Gough Map as wayfaring aid (something "a little unusual"); see also discussion of the Tarne Wathelyn, below.

39 For maps as "at least as much written as drawn," see discussion to chapter one, above. The Gough Map's text, by contrast, consists almost exclusively of labels giving settlement and topographic names, plus Roman numerals upon the road segments; exceptions will be discussed below. Parsons speculates that some text may have rubbed off in use, as a number of town-names have; longer discursive passages might be lost, for example in a lower corner where "information on the compiler . . . might usually be found";

pp. 3–4. For written legends on maps being more common where there is less geographical data available, see Carl Moreland and David Bannister, *Antique Maps* (1983), p. 40. For extended treatment of medieval maps' specific manner of combining textual and visual modes of representation, see final section to chapter one ("Painted Chamber *Compilatio*").

[40] On saltways see Hindle (1982) p. 212, Hilton p. 175, and Watkins pp. 22–23; nb. also discussion below. On Gough's roads into Droitwich (which incidentally appears only as . . . *wych* on the map) see Parsons p. 36; for area waterways, p. 34 but also Edwards and Hindle p. 131.

[41] See Hilton pp. 15–16, 175–76; those holding rights in the brine pits paid a duty to the exchequer and their shares were administered by borough officials.

[42] See Prestwich (1972) pp. 114–136 for victualling overall; pp. 123–125 provide tables for 1300, 1303, and 1304.

[43] Recall, incidentally, Hindle's account of Robert of Nottingham's itinerary through the midlands in order to procure wheat for the king in 1325—and the correspondence between the king's clerk's recorded distances for his trip and those on the Gough Map; see notes to chapter two.

[44] As Cuttino has emphasized, "it is the creation of the concept of the crown and the translation of that far-reaching concept into practice," especially on the part of kings' clerks, which seems to account for this development; p. 404. See also discussion and notes below.

[45] For an edition, see C. Bonnier, ed. "A List of 108 English Towns in the Fourteenth Century," *English Historical Review* 16 (1901) pp. 501–503; alternately, see H. Rothwell, ed., *English Historical Documents* 3, pp. 881–84; the phrase "clerk in his cups" appears in Bonnier's brief discussion. Translations/interpretations have been adapted in some cases from the handful of suggestions there provided but are otherwise my own. See Clanchy (1983) p. 251 for a brief notice. For a later English-Latin poem on the properties of seven English cities, in which some commentators have seen an affinity, see *A Manual of the Writings in Middle English*, vol. 7, p. 2238; see also "The Characteristics of the Counties," vol. 5, item #173.

[46] David Harvey (1969) p. 375. On a lack of coherence, see *Manual of Writings in Middle English* vol. 7, p. 2238; but nb. the allowance that "parts of the sequence suggest that the compiler had some plan of organization in mind."

[47] See Hilton pp. 198–99 overall. By contrast to highly specialized Droitwich, Hilton cites Worcester as "a perfect example of [economic] non-specialization"—appropriate therefore is this town's enrollment by the *List of 108 English Towns* not in economic or even institutional terms but for its "rhymers" or poets (*Rymeour de Wyrcestre* 58), an attribution extremely provocative insofar as it may relate to the fame of a network of West Midland vernacular lyric poets working in the early fourteenth century, some of whose poems appear in manuscripts such as London, B. L. Harley 2253. Nearby Hereford, like Worcester a center for the production of early Middle-English lyrics but another town largely non-specialized economically, has also apparently earned a national reputation for its secular love-poetry: hence the *Demayseles de Harford* ("Damsels of Hereford" 89). See G. L. Brook, *The Harley Lyrics* (Manchester, 1948).

[48] On Domesday Book and Chester's furriers, see Bonnier p. 503. For the Robin

Hood Ballads, see Barbara Hanawalt, "Ballads and Bandits," in Hanawalt (1992), pp. 154–175; 164 for the Sheriff of Nottingham.

[49] In 1324–25, a royal clerk named Robert of Nottingham took just such a trip, traveling in the region around the River Trent in Lincolnshire with an order to procure wheat for the king's wars—and with an early version of the Gough Map as his guide, apparently; he records the same mileage distances between the towns on his journey; see above notes. Also Hindle (1982) p. 198. For more on royal procurement see also pp. 201–202.

[50] See Davies p. 126.

[51] See Cuttino pp. 396, 404, 395; put most baldly, "These clerks, and officials like them, are propagandists for royal authority." See also discussion below.

[52] This region "has been noted for its red deer for centuries," says Parsons p. 29; see also p. 12 on "Gaick Forest, between Badenoch and Atholl, which is noted for its numerous herds of deer." See pp. 22, 23, 25 for references to Dean, Sherwood, etc.

[53] See Hilton pp. 15–16 on Arden as "not a legally defined entity as it would have been had it been a royal forest"; see Parsons pp. 8–9 on Arden's lack of icon.

[54] For forest cuts in Wales, see Prestwich (1972) pp. 110, 28; also Morris pp. 212, 223–25; see above note. For tables giving statistics on impressment see Rowlands pp. 62–63. On the widening of highways, see Guest p. 114.

[55] See Hilton pp. 13–16 in particular.

[56] See Richardson pp. 49–51.

[57] On the remarkable locational accuracy of the towns on the Gough Map, see Parsons p. 9.

[58] On map icons, see Woodward (1987) pp. 324–27, 354, 358. On status differentiation among Gough's settlement "vignettes" or "city-ideograms" see Parsons p. 9 and Schulz p. 448 but note also Hindle (1988)'s mildly dissenting view, p. 17. On a combination of symbols, see Moreland and Bannister p. 40.

[59] On the Hereford Map's Lincoln, see for example John Glenn, "Notes on the *Mappa Mundi* in Hereford Cathedral," pp. 60–63 in Ormrod (1986); also Harvey (1996) pp. 6–7. Gough's *Snowdonne* might constitute one exception to the map's generic practice of representing settlements; but the icon looks more likely to denote natural than human topography; see discussion above.

[60] On Calais and Gough, see Parsons p. 2; each of the other foreign sites noted above, incidentally—Dublin, Sluys, Bologne—held particular relevance in the reign of Edward I; see Powicke p. 668. See May McKisack, *The Fourteenth Century* (Oxford, 1959) pp. 148–49 for discussion regarding Edward III, for whom "the retention of Calais as an English possession" following its occupation in 1247 was a principle objective. Calais was also incidentally connected with the foundation of the Order of the Garter, formally inaugurated when Edward III "returned to England wearing the garland of Crécy and Calais" perhaps on St. George's Day, p. 251 and ff. For the Siege of Calais, pp. 135–37; on the Staple of Calais, pp. 353–54.

[61] See Schulz pp. 430, 468.

[62] Pierre de Langtoft, *Chronicle,* ed. T. Wright (London, 1868); see also Gransden (1974) pp. 470–86.

[63] On Gough's poor rendition of Scotland, see Parsons pp. 11–12; nb. Stones and

Simpson's point that in 1290 Edward I actually knew considerably less about Scotland than he did about the continent; see below notes for reference. All but two of the Scottish Earldoms are recorded by Gough, and the absence of those (e.g., Douglas) has been used to date an early version of the map at c.1300. On the much smaller ratio of district names to towns in England, see Parsons pp. 7–8. On Loch Tay, pp. 30, 12. On the greater likelihood of legendary material the further from a map's center of gravity, see Moreland and Bannister p. 40.

[64] For "acquisition of territory," see notes and discussion above. More importantly, for mention of another (later) map behind whose production there lay an evident "political intention"—Englishman John Harding's mid-fifteenth-century map of Scotland, surviving in several versions including B.L. MS Lansdowne 204, ff.226v-227r—see Delano-Smith and Kane, who suggest that the Harding map makes an explicit "attempt to persuade the English king to invade Scotland"; p. 32.

[65] The other lakes on Gough are Loch Tay in Perthshire; Windermere; and one with an indecipherable name in Wigtownshire; see Parsons pp. 30, 27. For the presence of the Tarne Wathelyn as "unusual" see Jewell p. 107; as Jewell notes, Parsons himself missed this identification ("I can find no trace of a lake of that name" p. 22)—apparently because he was being too literal-minded in his detective work.

[66] See Parsons pp. 25, 22 for his confident identifications of Glastonbury and Tintagel; both names are near indecipherable (e.g., just *gl*. . . for Glastonbury) but then, so are most towns in their vicinity.

[67] See Geoffrey of Monmouth I.11; on coming ashore, I.15. Totnes, of course, will serve not long afterwards as the source for Belinus's first and greatest road across the island, built to establish his coast-to-coast jurisdiction as king. See also the *Scema Britannie* (fig. 2.10) and its extended discussion in chapter two ("Map Genre Overlap: Form").

[68] See notes and discussion above (and mention below) on royal links to the *Historia*, including an early copy dedicated to King Stephen in 1136 and Edward I's presentation of another to Edward II in 1301. On Geoffrey as "the most popular work emanating from medieval Britain, and perhaps the most popular of all medieval histories," see Clanchy (1983) p. 127; likewise for "the credence he was given by reputable and scholarly writers."

[69] On "deference" and England's perceived myriad of advantages, see Davies pp. 6–10; on the general lack of "masterminding" to which however "the high-handedness and single-mindedness of Edward I" may serve as something of an exception, see Bartlett pp. 308–309 and Davies p. 24, 40; see also notes to chapter two. It may be to some extent misleading, Davies reflects, "to talk of a single policy or attitude on the part of the English crown towards the outlying countries of the British Isles or to detect sinister and far-reaching implications in every assertion of power by the king of England in Ireland, Scotland and Wales. Yet it cannot be denied that the power and attitude of the king of England was central, in one way or another, to the story of the Anglo-Norman and English domination of the British Isles. It could hardly be otherwise in a polity which was so monarchically dominated as that of medieval England"; p. 87. On Edward's responsibility, potentially, to press the English crown's claims to Scotland and Wales, see Davies p. 23; also Stones and Simpson (following note).

[70] For material related to the next several pages and to the Great Cause overall, see E. L. G. Stones and Grant G. Simpson, *Edward I and the Throne of Scotland* (Oxford, 1978), who provide a definitive account (vol. I) and collect most of the texts in question (vol. II). Stones and Simpson note that at the time of her death Margaret was "regarded by all as the rightful ruler," p. 270.

[71] See Stones and Simpson p. 12, who call this "a very parochial view"; nb. also Stone's comments in *Anglo-Scottish Relations, 1174–1328: Some Selected Documents* (London, 1965): "The assumption of some Scottish historians that Edward planned the subjugation of Scotland from the first is hardly borne out by the evidence," especially his inactivity in the crucial first months on the affair; p. xxv note 1.

[72] See Stones and Simpson pp. 21–23; see the *Calendar of Patent Rolls* for the late century generally on the correspondence between king and pope. Another significant event in this period was the November 28 1290 death at Lincoln of Edward's wife Eleanor of Castille. This caused him to put off dealing with Scottish matters until the beginning of March.

[73] The document recording this acknowledgment by the competitors, sometimes known as the "Award of Norham," is printed and translated by Stones, p. 56. See Stones and Simpson for the development of Edward's intervention, esp. p. 7: "We can only speculate about the stages by which the controversy in Scotland had come under Edward's jurisdiction between October 1290 and March 1291."

[74] On the great "mass of original documents" left behind by the affair, Stones and Simpson p. 2. The Great Rolls of John of Caen and Andrew de Tange are the most important. On Edward's correspondence with England's religious houses, see discussion below. See also Ralph A. Griffiths, *Conquerors and Conquered in Medieval Wales* (New York, 1994), chapter eleven ("Edward I, Scotland and the Chronicles of English Religious Houses") pp. 148–56.

[75] Stones and Simpson pp. 143, 147–48.

[76] Stones and Simpson pp. 137 and ff. esp. pp. 141, 144, 145, 148, 154. Overall Stones and Simpson characterize the affair as "a rather hasty improvisation" but also "enterprising and energetic," especially given that when asked for assistance, the monasteries "responded dutifully, but with no very great imagination."

[77] See previous notes on the *Historia* generally.

[78] Powicke pp. 683 and ff. for "dominated"; see Stone p. xxviii for the Bull, *Scimus Fili*; also Stones and Simpson pp. 154, 138 for discussion.

[79] See Stones and Simpson pp. 143, 155–56, who note that "it is perhaps curious that the trouble taken in 1300 to secure new historical information led, in the end, to nothing more than the mythological prologue, and the supplement for the period 1251–96." See Stones pp. 96–109 for the text.

[80] See Leckie overall on "the passage of dominion"; Stones and Simpson p. 156 for the Scottish skepticism and response; also Stones pp. 110–117. On the presentation to Edward II of a copy of the *Historia*, see discussion and notes above.

[81] For discussion of the manuscript, esp. ff.184v-186, and a full list of contents, see Stones and Simpson pp. 61–65; see also Vaughan (1958) pp. 241 and ff.

[82] Stones and Simpson, pp. 61–65.

[83] See Parsons p. 12; for the *regio montuosa* legend see Lewis (1987) p. 367. The

Corpus Christi version of Matthew's Map of Britain, incidentally, contains, like Gough, mention of the landing of Brutus—an inscription on Wales characterizing its inhabitants as a land of "busy and productive but bellicose men descended from the Trojan Brutus"; see Lewis pp. 369–71.

[84] On this warship's documentary value, see Parsons p. 2; pp. 12–13 for his description of the "stretched-out figure."

[85] Parsons p. 13 agrees that this is "an outline of a figure in a boat" but see note and discussion below for what I believe to be his far-fetched proposal regarding the figure's identification and significance.

[86] See Lewis (1987) pp. 212–215 for discussion of Henry's elephant; see Parsons p. 13 for a reprise of Sanders's proposal.

[87] See Parsons p. 13 for the prevailing generic speculation. On the *Chronica Majora* battle of sea monsters, see Lewis (1987) pp. 295–97; pp. 258–60 for the naval battle between Pisa and Genoa; and p. 297 for a possible link between these events. For the Mongol threat and the apocalyptic expectations this inaugurated, see Lewis (1987) pp. 103–105; incidentally, in "recapitulating the most significant events of the last half century" Paris "awarded pride of place to the invasions of the Tartars"; see also Powicke p. 110 note. Finally, note also Lewis's observation that there is a "recurrent pattern of significant coincidences between natural and human violence" in Matthew's *Chronicle*, but that this patter is "not consistently maintained throughout"; p. 297.

[88] See Powicke pp. 689, 694 on Welsh archers in Scotland; pp. 138, 590 on the years of Welsh subjugation. For Turberville's treason, see Freeman p. 445. Some prophetic literature of the period also makes reference to a combined Scottish and Welsh rising; see J. R. S. Phillips, "Edward II and the Prophets," in Ormrod (1986b), p. 191. See also the *Vita Edwardi Secundi*, p. 61.

[89] For *gladius*, see D. R. Howlett et al, *The Dictionary of Medieval Latin from British Sources* (Oxford, 1989), p. 1079.

[90] See Parsons p. 13 for Sanders's proposal and his own, which is for several reasons a far-fetched reading—not least, on the sobering grounds that the saga-material in question does not appear in any medieval manuscript but only in a centuries later addition of an entirely different narrative tone. But regarding *Orkneyinga Saga* as it stands, there are still problems. For some reason Parsons believes the Earl set out from Moray, a place treated in the saga's previous section but not that from which Rognvald sets out on his unlucky journey. This shipwreck, what is more, is not particularly ill-fated—no lives or even tempers are lost, the worst result being the loss of cargo and that the Earl gets cold. Once ashore, Rognvald takes the opportunity to compose a large number of verses and to go fishing with a local man, disguising himself in a cowl to hide his high status and then distributing his share of the catch to the local poor. Afterward he goes uneventfully home to Orkney. See chapter eighty-five, *Orkneyinga Saga: The History of the Earls of Orkney*, ed. and tr. Hermann Pálsson and Paul Edwards (London, 1978).

With the exception of Moray (a site of general importance and one whose place in the story and identification on the map may even be disputed), none of the sites associated with the story appears on the map—not even Shetland, despite the fact that these islands lay south of Norway, thus within the compass of the map. Nor does much knowledge of a Norse Orkney Saga on the part of any Westminster clerk or other user or copyist of the map seem very likely; *Orkneyinga Saga* was obscure even in Iceland, and if an English royal clerk may be supposed to have chosen, against all reason and likelihood, to add an Orkney detail or anecdote to this map, far more likely he would have been to choose something Arthurian, touching the four princes of Orkney so prominent at Arthur's court—Sir Gawain and his brothers. Going further, *Orkneyinga Saga* was never translated into Latin, Anglo-Norman, or Middle English. And finally there is also the matter of the shadowy drawing alongside the wreck, of which Parsons says: "it is possibly connected with illustrations of the legend of the earl's fishing expedition in disguise with a Dunrossness man." But Gough's shadow-image clearly shows one figure, not two—and the rectangle this figure pulls into his boat bears little resemblance to a net or fishing line.

[91] On the planned marriage between Margaret and Edward II as well as the plans made for Margaret's delivery to Scotland, see Powicke p. 599; also Stones and Simpson p. 204. The *Calendar of Patent Rolls* covering the late 1280s is full of documents relating to both these projects—marriage negotiations and crusade negotiations.

[92] See Stones and Simpson pp. 5 and ff. for a useful outline of events.

[93] For the text of Bishop Fraser's letter, see Stones and Simpson vol. II, p. 3; for its characterization as hasty and uncertain, pp. 4, 6; see also vol. I pp. 5–6 for discussion. On the confusion generated by events overall but especially this question of the Maid of Norway's death, see p. 4. In general, contemporary documents do not bother with the circumstances around Margaret's death (or perhaps do not have any information regarding this); none of the documents in Stones's *Anglo-Scottish Relations* or in Joseph Stevenson's *Illustrations of Scottish History* (Glasgow, 1834) goes beyond the simple *fact* of the Maid's death, to mention anything of its *manner*.

[94] As Swift said, "the metaphor be worn and stale, betwixt a state, and vessel in sail," *Imitations of Horace* Book I (1714); see for example T. B. Harbottle's *Anthology of Classical Quotations* (London, 1897), p. 381 and B. Stevenson's *The Home Book of Proverbs, Maxims and Familiar Phrases* (New York, 1948), p. 2209. In addition to those cited, perhaps relevant is Apuleius's use of *fortunae nautiagium*, "shipwrecked fortunes" or "a shipwreck of our fortunes."

[95] For brief overviews of *The Ship of State* and *The Death of Edward III*, see the *Manual of Writings in Middle English* vol. 5, p. 1488; another poem which employs the figure is *Richard the Redeless, or a Poem on the Deposition of Richard II* (c.1399)—also on the general subject of succession, although in this poem the ship of state does not comprise the work's central conceit as it does in the others.

[96] The Noble had been designed to replace the "ill-conceived and short-lived" Florin, issued only from January to August 1344, which had been based on the Flemish Florin, which "attained considerable circulation in northern Europe in the fourteenth century"; see Donald C. Baker, "Gold Coins in Mediaeval English Literature," *Speculum* 36 (1961), pp. 282–87, esp. pp. 283–4.

[97] See Baker esp. p. 284 note 12; p. 285 for "lord"/"record." For Edward III and the sea in general, see McKisack pp. 242–45; also Given-Wilson on the king's considerable expenditure on ships, p. 87; also p. 299, note 19.

[98] See Fredric L. Cheyette, "The Sovereign and the Pirates, 1332," *Speculum* 45 (1970), pp. 40–68, esp. pp. 45, 51–52, 54, 67. As Cheyette argues, such claims came to

be repeated frequently thereafter, until the assertion of an English "sovereignty of the sea" became "an integral part of the doctrine of royal sovereignty."

[99] See A. Z. Freeman, "A Moat Defensive: The Coast Defense Scheme of 1295," *Speculum* 42 (1967), pp. 442–462, esp. 442, 445. "England was beset by wars and the rumors of wars," Freeman says, and the fear of invasion was so acute that the king even "ordered foreign clerks removed from the seacoast and from access to navigable waterways near the sea."

[100] On the implications of English naval supremacy for the Scottish wars, see Prestwich (1972) p. 149.

[101] On the Bishop of St. Andrews's letter, see Stones and Simpson vol. II, p. 3 (text) and vol. I pp. 5–6 (discussion). The phrase "feared throughout Christendom" appears in the *Elegy on the Death of Edward I*; see also brief discussion above.

[102] See Stones p. 62 note; this clause was inserted into one exemplar of John of Caen's Great Roll and later included in all exemplars of Andrew de Tange's roll.

[103] See Cuttino p. 404; also J. R. Strayer, "The Laicization of French and English Society in the Thirteenth Century," *Speculum* 15 (1940), pp. 76–86.

Select Bibliography

PRIMARY SOURCES

An Anonymous Short English Metrical Chronicle. Ed. Ewald Zettl. EETS o.s. 196. London: Boydell & Brewer Ltd., 1935.

Baudri de Bourgueil. *Les oeuvres poétiques de Baudri de Bourgueil 1046–1130*. Ed. Phyllis Abrahams. Paris: Librarie Ancienne Honore Champion, 1926.

Brook, G. L. *The Harley Lyrics*. Manchester: Manchester University Press, 1948.

Butler, Thomas. *The Map off Ynglonnde*. New Haven, Beinecke MS 557, fols.47v-48r.

Calendar of Close Rolls, Henry III, 1227–1272. London: Public Record Office, 1902–38.

Calendar of Liberate Rolls, Henry III, 1216–40. London: Public Record Office, 1916.

Calendar of Patent Rolls, Edward I, 1272–81. London: Public Record Office, 1893–1901.

Chaucer, Geoffrey. *The Riverside Chaucer*. Ed. Larry D. Benson. 3rd Ed. Boston: Houghton Mifflin Co., 1987.

Crone, G. R., ed. *The World Map by Richard of Haldingham in Hereford Cathedral circa AD 1285*. London: Royal Geographical Society, 1954.

———. *Early Maps of the British Isles, AD 1000–AD 1579*. London: Royal Geographical Society, 1961.

Geoffrey of Monmouth. *The Historia Regum Britanniae of Geoffrey of Monmouth*. Ed. A. Griscom. London: Longman, Green & Co., 1929.

———. *Geoffrey of Monmouth's Historia Regum Britanniae: A Variant Version*. Ed. Jacob Hammer. Cambridge, MA: Mediaeval Academy of America, 1951.

———. *The History of the Kings of Britain*. Ed. and tr. Lewis Thorpe. Middlesex, 1987.

Giraldus Cambriensis. *Giraldus Cambrensis Opera*. Ed. J.S. Brewer. 8 vols. London: Longman, Green, Longman and Roberts, 1861–91.

———. *The Journey through Wales* and *The Description of Wales*. Ed. and tr. Lewis Thorpe. London: Penguin Classics, 1978.

Gervase of Canterbury. *The Historical Works of Gervase of Canterbury*. Ed. William Stubbs. 2 vols. London, 1879–80.

Gilson, J. P., ed. *Four Maps of Great Britain Designed by Matthew Paris About AD 1250.* London, 1928.

Henry of Huntingdon. *Historia Anglorum.* Ed. Thomas Arnold. Rolls Series. London: Rerum Britannicarum Medii Aevi Scriptores 74, 1879.

Higden, Ranulf. *Polychronicon.* Ed. J. R. Lumby. 9 vols. London: Longmans & Co., 1865–86.

Inquisitions and Assessments Relating to Feudal Aids, 1284–1431. London: York and Additions, Published by His Majesty's Stationery Office, 1899–1920

James, Henry, ed. *Facsimile of the Ancient Map of Great Britain in the Bodleian Library, Oxford, AD 1325–50.* Southampton, 1870. Reprinted 1973.

John, Trevor, ed. *The Warwickshire Hundred Rolls of 1279–80: Stoneleigh and Kineton Hundreds.* Oxford: Oxford University Press, 1992.

Langtoft, Peter. *The Chronicle of Pierre de Langtoft* . Ed. Thomas Wright. 2 vols. Rolls Series. London: Longsmans, Green, Reader and Dyer, 1868.

Layamon. *Brut.* Ed. G. L. Brook and R. F. Leslie. 2 vols. EETS 250, 277. London: Oxford University Press, 1963– 77.

———. *Brut.* Ed. and tr. Rosamund Allen. London: Dent, 1992.

Lehmann-Brockhaüs, Otto, ed. *Laateinische Schriftquellen zur kunst in England, Wales und Schottland, vom Jahre 901 biz zum Jahre 1307.* 5 vols. Munich: Prestel, 1955–60.

Liber Quotidianus Contrarotulatoris Garderobae. Ed. J. Nichol. Society of Antiquities. Oxford: Oxford University Press, 1787.

List of 108 English Towns in the Fourteenth Century. Ed C. Bonnier. *English Historical Review* 16 (1901): 501–503.

Mannyng, Robert. *Peter Langtoft's Chronicle (as illustrated and improv'd by Robert of Brunne).* Ed. Thomas Hearne. 2 vols. Oxford, 1725.

———. *The Story of England by Robert Manning of Brunne.* Ed. F. J. Furnivall. Rolls Series. 2 vols. London: Periodicals Service Company, 1887.

Matthew of Westminster. *Flowers of History.* Ed. and tr. C.D. Yonge. 2 vols. London: AMS Press, Incorporated, 1853

———. *Flores Historiarum.* Ed. H. R. Luard. 3 vols. London: Periodicals Service Company, 1890.

Moseley, C. W. R. D, ed. and tr. *The Travels of Sir John Mandeville.* Middlesex: Penguin Books, 1983.

Notitia Dignitatum. Ed. Otto Seeck. 1876, repr. Frankfurt am Main. Minerva, 1962.

Orkneyinga Saga. Ed. Finnbogi Gudmundsson. *Íslenzk fornrit,* vol. 34. Reykjavík: Íslenzka fornritafélag, 1965.

Orkneyinga Saga: The History of the Earls of Orkney. Ed. and tr. Hermann Pálsson and Paul Edwards. London: Viking Penguin, 1978.

Paris, Matthew. *Chronica Majora.* Ed. H. R. Luard. Rolls Series. 7 vols. London, 1872–83.

———. *The Illustrated Chronicles of Matthew Paris: Observations of Thirteenth-Century Life.* Ed. Richard Vaughan. Cambridge: Alan Sutton, 1993.

Parsons, E. J. S., ed. *The Map of Great Britain ca. AD 1360 known as the Gough Map.* Oxford: Bodleian Library, 1958.

Polo, Marco. *The Travels.* Ed. and tr. Ronald Latham. Harmondsworth: Penguin, 1979.

Robert of Gloucester. *The Metrical Chronicle of Robert of Gloucester.* Ed. W. A. Wright. Rolls Series. 2 vols. London: Printed for Her Majesty's Stationery Office, 1887.

Roger of Hoveden. *Annals of Roger de Hoveden.* In *Leges Anglo-Saxonicae,* ed. C. S. Wilkins. London, 1721.

Roger of Hoveden. *Annals of Roger de Hoveden.* Ed. and tr. H. T. Riley. London: H. G. Bohn, 1853.

———. *Chronica.* Ed. W. Stubbs. Rolls Series. 4 vols. London: Longman, 1868–71.

———. *Flores Historiarum.* Ed. H. G. Hewlett. Rolls Series. 3 vols. London: Periodicals Service Company, 1886–89.

Seymour, M. C., ed. *The Metrical Version of Mandeville's Travels.* EETS o.s. 269. London: Oxford University Press, 1973.

Stevenson. Joseph, ed. *Illustrations of Scottish History from the Twelfth to the Sixteenth Century.* Glasgow: AMS Press, Incorporated, 1834.

Stones, E. L. G., ed. *Anglo-Scottish Relations 1174–1328: Some Selected Documents.* London: Oxford University Press, 1965.

Stones, E. L. G. and Grant G. Simpson, eds. *Edward I and the Throne of Scotland, 1290–1296.* Vol. II: Texts. Oxford: Oxford University Press, 1978.

Thorpe, Lewis, ed. and tr. *Two Lives of Charlemagne.* Middlesex: Pearson Higher Education, 1969.

Vita Edwardi Secundi/The Life of Edward the Second. Ed. and tr. N. Denholm-Young. London: Thomas Nelson and Sons, Ltd., 1957.

Wace. *Le Roman de Brut de Wace.* Ed. Ivor Arnold. Paris: SATF, 1938.

Walter of Coventry. *Memoriale Fratris Walteri de Coventria.* Ed. W.M. Stubbs. Rolls Series. 2 vols. London: Longman and Co., 1872–3.

Webb, John, ed. *Roll of the Household Expenses of Richard de Swinfield, Bishop of Hereford, during part of the years 1289 and 1290.* Camden Society. 2 vols. London: AMS, 1854–55.

Wright, Thomas, ed. and tr. *The Political Songs of England, from the Reign of John to that of Edward II.* London: Camden Society, 1839.

SECONDARY SOURCES

Abulafia, David. *Frederick II: A Medieval Emperor.* New York: Oxford University Press, 1992.

———. "Ethnic Variety and Its Implications: Frederick II's Relations with Jews and Muslims." In Tronzo (1994), 213–224.

Ahmad, Nafis. *Muslims and the Science of Geography.* Dacca: University Press, 1980.

Alexander, J. J. G. and Paul Binski, eds. *Age of Chivalry: Art in Plantagenet England, 1200–1400.* London: Royal Academy of Arts, 1987.

———. "The Illustrated Manuscripts of the *Notitita Dignitatum.*" In Goodburn and Bartholomew (1976), 11–25.

——— and M. T. Gibson, eds. *Medieval Learning and Literature: Essays Presented to R. W. Hunt*. Oxford: Oxford University Press, 1976.

Alington, Gabriel. *The Hereford* Mappamundi. Hereford: Gracewing, 1996.

Allen, Rosamund. Introduction. *Lawman's* Brut. London: J. M. Dent, 1992.

Anderson, Benedict. *Imagined Communities: Reflections on the Origins and Spread of Nationalism*. London: Verso, 1983.

Arnheim, Rudolf. *The Power of the Center: A Study of Composition in the Visual Arts*. Berkeley: University of California Press, 1982.

Ashcroft, Bill, Gareth Griffiths, and Helen Tiffen. *The Empire Writes Back: Theory and Practice in Postcolonial Literatures*. London: Routledge, 1989.

Backhouse, Janet. "'A Very Old Book': The Burdett Psalter-Hours, Made for a Thirteenth-Century Hospitaller." In Cassidy and Wright (2000), 55–66.

Bagrow, Leo. *History of Cartography*. Revised R. A. Skelton. Chicago: Precedent Publishing, 1985 [1964].

Bahat, Daniel. *The Illustrated Atlas of Jerusalem*. Tr. Shlomo Ketko. New York: Macmillan Reference Books, 1990.

Bailey, M. "The *Mappa mundi* Triptych: The Full Story of the Hereford Cathedral Panels." *Apollo* 137 (1993): 374–78.

Baker, Donald C. "Gold Coins in Medieval English Literature." *Speculum* 36 (1961): 282–87.

Barber, Peter. "Old Encounters New: The Aslake World Map." In Pelletier (1989), 69–88.

———. "England I: Pageantry, Defense and Government: Maps at Court." In Buisseret (1992), 26–56.

———. "The Evesham World Map: A Late Medieval English View of God and the World." *Imago Mundi* 47 (1995): 13–33.

——— and M. Brown. "The Aslake World Map." *Imago Mundi* 44 (1992), **pp to come**.

Barker, Felix and Peter Jackson. *The History of London in Maps*. London: Barrie and Jenkins, 1990.

Barnes, Trevor J. and James S. Duncan, eds. *Writing Worlds: Discourse, Text, and Metaphor in the Representation of Landscape*. New York: Routledge, 1992.

Barraclough, Geoffrey, ed. and tr. *Medieval Germany, 911–1250: Essays by German Historians*. Oxford: Basil Blackwell, 1938.

———. *The Origins of Modern Germany*. 3rd ed. Oxford: Basil Blackwell, 1988 [1947].

Barrow, G. W. S. *The Anglo-Norman Era in Scottish History*. Oxford: Oxford University Press, 1980.

Bartlett, Robert. *The Making of Europe: Conquest, Colonization and Cultural Change*. Princeton, NJ: Allen Lane, 1993.

Baswell, Christopher. "Talking Back to the Text: Marginal Voices in Medieval English Literature." In Morse, Doob and Allen (1992), 121–60.

Beazley, Charles Raymond. *The Dawn of Modern Geography: A History of Exploration and Geographical Science from the Conversion of the Roman Empire to AD 900*. 3 vols. Cambridge: Cambridge University Press, 1897–1906.

———. "New Light on Some Medieval Maps." *Geographical Journal* 15 (1900): 135–38; *Geographical Journal* 16 (1900), 322.

Beckwith, Sarah. *Christ's Body: Identity, Culture, and Society in Late Medieval Writings.* New York: Routledge, 1993.

Bennett, Adelaide. "A Late Thirteenth-Century Psalter-Hours from London." In Ormrod (1986a) 15–30.

Bennett, Michael J. *Community, Class, and Careerism: Cheshire and Lancashire Society in the Age of Sir Gawain and the Green Knight.* Cambridge: Cambridge University Press, 1983.

Bevan, W. L. and H. W. Phillott. *Mediaeval Geography, an Essay in Illustration of the Hereford Mappa Mundi.* London, 1873.

Biddick, Kathleen. "Genders, Bodies, Borders: Technologies of the Visible." *Speculum* 68 (1993), 389–418.

Biddle, Martin, ed. *King Arthur's Round Table: An Archaeological Investigation.* Woodbridge, Suffolk: Boydell Press, 2000.

Binski, Paul. *The Painted Chamber at Westminster Palace.* London: Society of Antiquaries of London, 1986.

———. *Westminster Abbey and the Plantagenets: Kingship and the Representation of Power.* New Haven: Yale University Press, 1995.

Bloch, Marc. *Feudal Society.* Tr. L. A. Manyon. 2 vols. Chicago: University of Chicago Press, 1963.

Blunt, Alison and Gillian Rose, eds. *Writing Women and Space: Colonial and Postcolonial Geographies.* London: Guilford Publications, Incorporated, 1994.

Borenius, Tancred. "The Cycle of Images in the Palaces and Castles of Henry III." *Journal of the Warburg and Courtauld Institutes* 6 (1943): 47–50.

Bourdieu, Pierre. *Language and Symbolic Power.* Tr. Gino Raymond and Matthew Adamson. Cambridge, MA: Harvard University Press, 1991.

Brackmann, Albert. "The Beginning of the National State in Medieval Germany and the Norman Monarchies." In Barraclough (1938).

Britain Before the Norman Conquest. Map. Ordnance Survey. Southampton, 1973.

Brown, Lloyd A. *The Story of Maps.* Boston: Little, Brown and Co., 1949.

Brown, Peter S. *The Cult of the Saints: Its Rise and Function in Latin Christianity.* Chicago: University of Chicago Press, 1981.

Buisseret, David, ed. *Monarchs, Ministers and Maps: The Emergence of Cartography as a Tool of Government in Early Modern Europe.* Chicago: University of Chicago Press, 1992.

Bunson, Matthew. *Encyclopedia of the Middle Ages.* New York, 1995.

Cameron, Kenneth. English Place-Names. London: B. T. Batsford, 1961.

Camille, Michael. "Illustrations in Harley MS 3487 and the Perception of Aristotle's *Libri naturales* in Thirteenth-Century England." In Ormrod (1986a), 31–44.

———. *Image on the Edge: The Margins of Medieval Art.* Cambridge, MA: Harvard University Press, 1992.

———. "The Dissenting Image: A Postcard from Matthew Paris." In Copeland (1996)

115–150.

Campbell, Mary B. *The Witness and the Other World: Exotic European Travel Writing, 400–1600.* Ithaca, NY: Cornell University Press, 1988.

Campbell, Tony. "Portolan Charts from the Late Thirteenth Century to 1500." In Harley and Woodward (1987), 371–463.

Canning, Joseph. *A History of Medieval Political Thought 300–1450.* London: Routledge, 1996.

Cantor, Norman. *Inventing the Middle Ages: The Lives, Works, and Ideas of the Great Medievalists of the Twentieth Century.* New York: William Morrow & Co., 1991.

Carr, A. D. *Medieval Anglesey.* Llangefni: Anglesey Antiquarian Society, 1982.

———. *Medieval Wales.* New York: St. Martin's Press, 1995.

Carruthers, Mary. *The Book of Memory: A Study of Memory in Medieval Culture.* Cambridge: Cambridge University Press, 1979.

Cassidy, Brendan and Rosemary Muir Wright, eds. *Studies in the Illustration of the Psalter.* Stamford, Lincolnshire: Tyas, 2000.

Cheyette, Fredric L. "The Sovereign and the Pirates, 1332." *Speculum* 45 (1970): 40–68.

Chibnall, Marjorie. *Anglo-Norman England, 1066–1166.* Oxford: Basil Blackwell, 1986.

Clanchy, M. T. *England and Its Rulers, 1066–1272.* Oxford: Basil Blackwell, 1983.

———. *From Memory to Written Record: England, 1066–1307.* 2nd ed. Oxford: Basil Blackwell, 1993.

———. "England in the Thirteenth Century: Power and Knowledge." In Ormrod (1986a), 1–14.

Codrington, Thomas. *Roman Roads in Britain.* London: Society for Promoting Christian Knowledge, 1903.

Colvin, H. M. *et al.*, eds. *The History of the King's Works: The Middle Ages.* London: Her Majesty's Stationery Office, 1963.

Copeland, Rita. *Rhetoric, Hermeneutics and Translation in the Middle Ages: Academic Traditions and Vernacular Texts.* Cambridge: Cambridge University Press, 1991.

———, ed. *Criticism and Dissent in the Middle Ages.* Cambridge: Cambridge University Press, 1996.

Cormack, Leslie B. "Good Fences Make Good Neighbors: Geography as Self-Definition in Early Modern England." *Isis* 82 (1991): 639–661.

Cosgrove, Denis and Stephen Daniels, eds. *The Iconography of Landscape: Essays in the Symbolic Representation, Design and Use of Past Environments.* Cambridge: Cambridge University Press, 1988.

Crane, Susan. *Insular Romance: Politics, Faith and Culture in Anglo-Norman and Middle English Literature.* Berkeley: University of California Press, 1986.

Crone, G. R. *The Hereford World Map.* London: Royal Geographical Society, 1949.

———. *Maps and Their Makers.* London: Hutchinson's University Library, 1953.

———. Introduction. *The World Map by Richard of Haldingham in Hereford Cathedral circa AD 1285.* London: Royal Geographical Society, 1954.

———. Introduction. *Early Maps of the British Isles AD 1000–AD 1579.* London: Royal Geographical Society, 1961.

———. "New Light on the Hereford Map." *Geographical Journal* 131 (1965): 447–62.

Cross, P. R. and S. D. Lloyd, eds. *Thirteenth-Century England IV. Proceedings of the Newcastle-upon-Tyne Conference 1991*. Woodbridge, Suffolk: Boydell, 1992.

Curtius, Ernst Robert. *European Literature and the Latin Middle Ages*. Tr. William R. Trask. Princeton: Princeton University Press, 1973.

Cuttino, G. P. "King's Clerks and the Community of the Realm." *Speculum* 29 (1954): 395–409.

Darby, H. C., ed. *A New Historical Geography of England*. Cambridge: Cambridge University Press, 1973.

———, ed. *A New Historical Geography of England Before 1600*. Cambridge: Cambridge University Press, 1976

Davies, R. R. "Edward I and Wales." In Herbert and Jones (1988), 1–10.

———. *Domination and Conquest: The Experience of Ireland, Scotland and Wales, 1100–1300*. Cambridge: Cambridge University Press, 1990.

de Certeau, Michel. *The Practice of Everyday Life*. Tr. Steven Rendall. Berkeley: University of California Press, 1984.

———. *The Writing of History*. Tr. Tom Conley. New York: Columbia University Press, 1988.

de Hamel, Christopher. "*Mappa Mundi*: The Hereford Map." Sotheby's Pamphlet. London, 1990.

Delano-Smith, Catherine and Roger J. P. Kain. *English Maps: A History*. London: British Library, 1999.

——— and Elizabeth Morley. *Maps in Bibles, 1500–1600: An Illustrated Catalog*. Geneva: Genève, 1991.

Denholm-Young, N. "The Tournament in the Thirteenth Century." In *Studies in Medieval History Presented to F. M. Powicke*, ed. R. W. Hunt, W. A. Pantin, and R. W. Southern, 240–248. Oxford: Clarendon Press, 1948

. "The *Mappa Mundi* of Richard of Haldingham at Hereford." *Speculum* 32 (1957): 307–314.

Destombes, Marcel, ed. *Mappemondes, AD 1200–1500*. Amsterdam: Israel, 1964.

Dilke, O. A. W. "Maps in the Service of the State: Roman Cartography to the End of the Augustan Era." In Harley and Woodward (1987), 201–211.

———. "Roman Large-Scale Mapping in the Early Empire." In Harley and Woodward (1987), 212–233.

———. "Itineraries and Geographical Maps in the Early and Late Roman Empires." In Harley and Woodward (1987), 234–257.

Dingley, Thomas. *History From Marble, compiled in the reign of Charles II by Thomas Dingley, gent.* Ed. J. G. Nichols. Camden Society. London, 1867 8.

Duffy, Eamon. *The Stripping of the Altars: Traditional Religion in England, 1400–1580*. New Haven: Yale University Press, 1992.

Edson, Evelyn. *Mapping Time and Space: How Medieval Mapmakers Viewed Their World*. London: The British Library, 1997.

Edwards, James Frederick and Brian Paul Hindle. "The Transportation System of

Medieval England and Wales." *Journal of Historical Geography* 17:2 (1991): 123–34.

Egerton, Samuel Y., Jr. "From Mental Matrix to *Mappamundi* to Christian Empire: The Heritage of Ptolemaic Cartography in the Renaissance." In Woodward (1987), 10–50.

Emden, A. B. *A Biographical Register of the University of Oxford to AD 1500.* 3 vols. Oxford: Oxford University Press, 1957.

Estey, F. N. "Charlemagne's Silver Celestial Table." *Speculum* 18 (1943): 112–117.

Fabian, Johannes. *Time and the Other: How Anthropology Makes Its Object.* New York: Columbia University Press, 1983.

Field, P. J. C. "The Winchester Round Table." *Notes and Queries* 25:3 (1978): 204.

Flint, Valerie I. J. *The Imaginative Landscape of Christopher Columbus.* Princeton: Princeton University Press, 1992.

———."The Hereford Map: Its Author(s), Two Scenes and a Border." *Transactions of the Royal Historical Society*, 6th series, 8 (1998): 19–44.

Figgis, John Neville. *The Divine Right of Kings.* New York: Harper Torchbooks, Harper & Row, Publishers, 1896.

Fox-Davies, A.C. *A Complete Guide to Heraldry.* Revised J. P. Brooke. London: Nelson, 1969 [1909].

Foucault, Michel. *The Archaeology of Knowledge.* Tr. A. M. Sheridan-Smith. New York: Pantheon Books, 1972.

———. *Discipline and Punish: The Birth of the Prison.* Tr. Alan Sheridan. New York: Vintage Books, 1979.

———. "Of Other Spaces." *Diacritics* 16 (1986): 22– 25.

Franklyn, Julian and John Tanner. *An Encyclopedic Dictionary of Heraldry.* London: Pergamon, 1970.

Freeman, A. Z. "A Moat Defensive: The Coast Defense Scheme of 1295." *Speculum* 42 (1967): 442–62.

Freiesleben, H. C. "The Origin of Portolan Charts." *Journal of Navigation* 37 (1984): 194–199.

Friedman, John Block. *The Monstrous Races in Medieval Art and Thought.* Cambridge, MA : Harvard University Press, 1981.

Frugnoni, Chiara. *A Distant City: Images of Urban Experience in the Medieval World.* Princeton: Princeton University Press, 1991.

Fuhrmann, Horst. *Germany in the High Middle Ages c.1050–1200.* Tr. Timothy Reuter. Cambridge: Cambridge University Press, 1986.

Galbraith, V. H. "Good and Bad Kings in Medieval English History." *History* 30 (1945): 119–132.

Gale, Roger. "An Essay Towards the Recovery of the Courses of the Four Great Roman Ways." In Leland (1770), vol. 6, 116–150.

Gilbert, Martin. *British History Atlas.* London: MacMillan Publishing, 1969.

Gilson, J. P. and H. Poole, eds. *Four Maps of Great Britain Designed by Matthew Paris About AD 1250.* London, 1928.

Given-Wilson, Chris. *The Royal Household and the King's Affinity: Service, Politics, and*

Finance in England, 1360–1413. New Haven: Yale University Press, 1986.
Glanville, Phillipa. *London in Maps*. London: The Connoisseur, 1972.
Glasscock, R. E. "England Circa 1334." In Darby (1976), 136–85.
Glenn, John. "Notes on the *Mappa Mundi* in Hereford Cathedral." In Ormrod (1986a), 60–63.
Goodburn, Roger and Philip Bartholomew, eds. *Aspects of the Notitia Dignitatum*. Oxford: British Archaeological Reports International Series, 1976.
Gough, H. *Itinerary of King Edward the First*. 2 vols. Paisley: List and Index Society, 1900.
Gough, H. and J. Parker. *A Glossary of Terms Used in Heraldry*. Oxford: James Parker & Co., 1894.
Gough, R. *British Topography*. 2 vols. London: T. Payne & J. Nichols, 1780.
Gransden, Antonia. *Historical Writing in England vol. I, c.550 to c.1307*. London: Routledge, 1974.
———. *Historical Writing in England vol. II, c.1307 to the Early Sixteenth Century*. London: Routledge, 1982.
Greenblatt, Stephen. *Marvelous Possessions: The Wonder of the New World*. Chicago: University of Chicago Press, 1991.
Gregory, Derek. *Geographical Imaginations*. Cambridge, MA: Blackwell Publishers, 1994.
Griffiths, Ralph A. *Conquerors and Conquered in Medieval Wales*. New York: Palgrave Macmillan, 1994.
Guest, Edwin. "The Four Roman Ways." *The Archaeological Journal* 14 (1857) 99–118.
Gurevich, Aron. *Categories of Medieval Culture*. London: Routledge Kegan & Paul, 1985.
Habermas, Jurgen. *The Structural Transformation of the Public Sphere: An Inquiry into a Category of Bourgeois Society*. Tr. Thomas Burger. Cambridge MA: Blackwell Publishers, 1989.
———. "Further Reflections on the Public Sphere." In *Habermas and the Public Sphere*, ed. Craig Calhoun, 425–61. Cambridge, MA: MIT Press, 1989.
Hahn, Thomas. "Imagining the Exotic: Visual Representations of "Indians" in Late Medieval Vernacular Manuscripts." Paper delivered at the University of Minnesota, "Strangers in Medieval Society" Conference. February 1994.
Hall, D. J. *English Mediaeval Pilgrimage*. London: Routledge Kegan & Paul, 1965.
Hallam, Elizabeth, ed. *The Plantagenet Encyclopedia*. London: Tiger Books International, 1996.
Hampe, Karl. *Germany Under the Salian and Hohenstaufen Emperors*. Tr. Ralph Bennett. Cambridge: Blackwell Publishers, 1986.
Hanawalt, Barbara, ed. *Chaucer's England: Literature in Historical Context*. Minneapolis: University of Minnesota Press, 1992.
———. "Ballads and Bandits." In Hanawalt (1992), 154–175.
Hanna, Ralph III. "Sir Thomas Berkeley and his Patronage." *Speculum* 64 (1989): 878–916.

Hanna, Ralph III. "*Compilatio* and the Wife of Bath: Latin Backgrounds, Ricardan Texts." In Minnis (1989), 1–11.

Harbottle, T. B. *Anthology of Classical Quotations.* London, 1897.

Harley, J. B. "The Evaluation of Early Maps: Towards a Methodology." *Imago Mundi* 22 (1970): 62–74.

———. "The Iconology of Early Maps." In *Imago et Mensura Mundi: Atti del IX Congresso Internazionale di Storia della Cartografia.* Ed. Carla Clivio Marzoli, 29–38. Pisa, 1981.

———. "The Map and the Development of the History of Cartography." In Harley and Woodward (1987), 1–42.

———. "Maps, Knowledge and Power." In Cosgrove and Daniels (1988), 277–312.

———. "Deconstructing the Map." In Barnes and Duncan (1992), 231–247.

———. "The Evaluation of Early Maps." In "History of Cartography Typescript." Hampshire Record Office. Winchester, no date: 203–220.

Harley, J. B. and David Woodward, eds. *The History of Cartography Volume I: Cartography in Prehistoric, Ancient, and Medieval Europe and the Mediterranean.* Chicago: University of Chicago Press, 1987.

Harvey, David. *Explanation in Geography.* London: Edward Arnold, 1969.

———. *The Condition of Postmodernism: An Enquiry into the Conditions of Cultural Change.* Oxford: Blackwell Publishers, 1989.

Harvey, P. D. A. "Medieval Maps: An Introduction." In Harley and Woodward (1987), 283–285,

———. "Local and Regional Cartography in Medieval Europe." In Harley and Woodward (1987), 464–501.

———. *Medieval Maps.* Toronto: University of Toronto Press, 1991.

———. "Matthew Paris's Maps of Britain." In Cross and Lloyd (1992).

———. *Maps in Tudor England.* Chicago: University of Chicago Press, 1993.

———. *Mappa Mundi: The Hereford World Map.* London: Hereford Cathedral and the British Library, 1996.

———. "The Sawley Map and Other World Maps in Twelfth-Century England." *Imago Mundi* 49 (1997): 33–42.

———. "A Thirteenth-Century Plan for Waltham Abbey, Essex." In "History of Cartography Typescript." Hampshire Record Office. Winchester, no date: 1–6.

Haslam, Graham. "The Duchy of Cornwall Map Fragment." In Pelletier (1989) 33–44.

Hassall, M. W. C. "Britain and the *Notitia.*" In Goodburn and Bartholomew (1976), 103–118.

Haverkamp, Alfred. *Medieval Germany, 1056–1273.* Tr. Helga Braun and Richard Mortimer. Oxford: Oxford University Press, 1988.

Heer, Friedrich. *The Holy Roman Empire.* Tr. Janet Sondheimer. New York: Frederick A. Praeger, 1968.

Herben, Stephen J., Jr. "The Vercelli Book: A New Hypothesis." *Speculum* 10 (1935): 91–94

Herbert, Trevor and Gareth Elwyn Jones, eds. *Edward I and Wales.* Cardiff: University of

Wales Press, 1988.

Herde, Peter. "Literary Activities of the Imperial and Papal Chanceries during the Struggle between Frederick II and the Papacy." In Tronzo (1994), 227–239.

Hexter, Ralph. "*Latinitas* in the Middle Ages: Horizons and Perspectives." *Helios* 14 (1987): 69–93.

Higgins, Iain Macleod. "Defining the Earth's Center in a Medieval 'Multi-Text': Jerusalem in *The Book of John Mandeville*." In Tomasch (1998), 29–53.

Hilton, R. H. *A Medieval Society: The West Midlands at the End of the Thirteenth Century.* London: Weidenfeld & Nicolson, 1966.

Hindle, B. P. "The Road Network of Medieval England and Wales." *Journal of Historical Geography* 2:3 (1976): 207–221.

———. "The Towns and Roads of the Gough Map." *The Manchester Geographer* 1 (1980): 35–40.

———. "Roads and Tracks." In *The English Medieval Landscape*, ed. Leonard Cantor, 193–217. Philadelphia: University of Pennsylvania Press, 1982.

———. *Maps for Local History.* London: B.T. Batsford, 1988.

Holbrook, Sue Ellen. "Malory's Identification of Camelot as Winchester." In *Studies in Malory*, ed. James W. Spisak, 13–27. Kalamazoo: Medieval Institute Publications, 1985.

Hornblower, Simon and Antony Spawforth, eds. *The Oxford Classical Dictionary*. 3rd ed. Oxford: Oxford University Press, 1996

Howard, Donald R. *Writers and Pilgrims: Medieval Pilgrimage Narratives and Their Posterity.* Berkeley: University of California Press, 1980.

Howlett, D. R. *et al. The Dictionary of Medieval Latin from British Sources.* Oxford: Oxford University Press, 1989.

Hucker, Bernd-Ülrich. *Kaiser Otto IV.* Hannover: Hahnsche Buchh, 1990.

Huizinga, Johan. *The Waning of the Middle Ages: A Study of the Forms of Life.* Tr. F. J. Hopman. New York: St. Martin's Press, 1924.

Irvine, Martin. "'Bothe Text and Gloss': Manuscript Form, the Textuality of Commentary, and Chaucer's Dream Poems." In Morse, Doob and Woods (1992), 81–119.

Jack, R. Ian. *Medieval Wales.* Ithaca, NY: Cornell University Press, 1972.

James, M. H. *Marvels of the East.* Roxburghe Club. Oxford: University Press, 1929.

James, Thomas Beaumont. *The Palaces of Medieval England c. 1050–1550.* London: B.A. Seaby, Ltd., 1990.

Jancey, Meryl. *Mappa Mundi: The Map of the World at Hereford Cathedral.* Hereford: The Friends of the Hereford Cathedral, 1987.

Jeffrey, David Lyle, ed. *A Dictionary of Biblical Tradition in English Literature.* Grand Rapids, MI: William B. Eerdmans Publishing Company, 1992.

Jervis, W. W. *The World in Maps: A Study in Map Evolution.* London: George Philip and Son, 1936.

Jewell, Helen. *The North-South Divide: The Origins of Northern Consciousness in England.* Manchester: Manchester University Press, 1994.

John, Trevor. Introduction. *The Warwickshire Hundred Rolls of 1279–80.* Oxford: Oxford University Press, 1992.

Jusserand, J. J. *English Wayfaring Life in the Middle Ages.* Tr. L.T. Smith. London: Milford House, 1889.

Kantorowicz, Ernst. *The King's Two Bodies: A Study in Medieval Political Theology.* Princeton, NJ: Princeton University Press, 1957.

Kaufmann, Virginia Roehrig. "The Magdeburg Rider: An Aspect of the Reception of Frederick II's Roman Revival North of the Alps." In Tronzo (1994), 63–88.

———. "Creatures from the Ends of the Earth at the Guelph Court in Braunschweig." Paper delivered at the University of Minnesota, "Strangers in Medieval Society" Conference. February 1994.

Kenyon, John R. and Richard Avent, eds. *Castles in Wales and the Marches: Essays in Honour of D. J. Cathcart King.* Cardiff: University of Wales Press, 1987.

Kimble, George H. T. *Geography in the Middle Ages.* London: Methuen & Co. Ltd., 1938.

Kitzinger, Ernst. "World Map and Fortune's Wheel: A Medieval Mosaic Floor in Turin." *Proceedings of the American Philosophical Society* 117: 5 (1973): 344–373.

Kline, Naomi Reed. *Maps of Medieval Thought: The Hereford Paradigm.* Woodbridge, Suffolk: Boydell Press, 2001

Knowles, Dom. David. "The *Mappa Mundi* of Gervase of Canterbury." *The Downside Review* 48 (1930): 237–247.

Köhler, Erich. *L'aventure chevaleresque: Idéal et réalité dans le roman courtois.* Paris: Gallimard, 1974.

Kupfer, Marcia. "The Lost *Mappa Mundi* at Chalivoy-Milon." *Speculum* 66 (1991): 540–71.

———. "Medieval World Maps: Embedded Images, Interpretive Frames." *Word and Image* 10 (1994): 262–88.

Lacy, N. J., ed. *The New Arthurian Encyclopedia.* New York: Garland, 1996.

Lanman, Jonathan T. *On the Origin of Portolan Charts.* Chicago: The Newberry Library, 1987.

———. *Glimpses of History from Old Maps: A Collector's View.* Tring, Herts.: Map Collector Publications, 1989.

Latham, R.E. *Revised Medieval Latin Word-List from British and Irish Sources.* London: Oxford University Press, 1965.

Lawton, David. "Analytical Survey I: Literary History and Cultural Study." *New Medieval Literatures* 1 (1997): 237–269.

———. "The Surveying Subject and the 'Whole World' of Belief: Three Case Studies." *New Medieval Literatures* 4 (2000): 9–37.

Leckie, R. William. *The Passage of Dominion: Geoffrey of Monmouth and the Periodization of Insular History in the Twelfth Century.* Toronto: University of Toronto Press, 1981.

Lefebvre, Henri. *The Production of Space.* Tr. Donald Nicholson-Smith. Oxford: Blackwell Publishers, 1991.

Legge, M. Dominica. *Anglo-Norman Literature and its Background.* Oxford: Greenwood

Publishing Group, Inc., 1963.

Leland, John. *The Itinerary of John Leland the Antiquary, published by Mr. Thomas Hearne.* 6 vols. Oxford, 1770.

Letts, Malcolm. "The Pictures in the Hereford *Mappa Mundi.*" *Notes and Queries* 200 (1955): 2–6.

———. *The Pictures in the Hereford Mappa Mundi.* Hereford, 1955.

Lewis, Suzanne. *The Art of Matthew Paris in the Chronica Majora.* London: University of California Press, 1987.

———. *Reading Images: Narrative Discourse and Reception in the Thirteenth-Century Illuminated Apocalypse.* New York: Cambridge University Press, 1995.

Lindemann, Rolf. "A New Dating of the Ebsdorf *Mappamundi.*" In Pelletier (1989), 45–49.

Loomis, R. S. "Edward I, Arthurian Enthusiast." *Speculum* 28 (1953): 114–127.

———, ed. *Arthurian Literature in the Middle Ages.* Oxford: Clarendon Press, 1959.

Lynam, Edward. *British Maps and Map Makers.* London: William Collins, 1944.

———. *The Mapmaker's Art: Essays on the History of Maps.* London: Batchworth Press, 1953.

A Manual of the Writings in Middle English, 1050–1500. 9 vols. New Haven: Yale University Press, 1967–1998.

Margary, Ivan. *Roman Roads in Britain.* London: Phoenix House, 1955.

Marshall, M. H. "Thirteenth-Century Culture as Illustrated by Matthew Paris." *Speculum* 14 (1939): 465–77.

Marvin, William P. "*Ritus Venandi:* Discourses of the Medieval English Hunt." Diss. U of Minnesota, 1998.

Massey, Doreen. *Space, Place, and Gender.* Minneapolis: University of Minnesota, 1994.

McKitterick, Rosamund. *The Carolingians and the Written Word.* Cambridge: Cambridge University Press, 1989.

McKisack, May. *The Fourteenth Century.* Oxford: Clarendon Press, 1959.

Meisel, Janet. *Barons of the Welsh Frontier.* Lincoln: University of Nebraska Press, 1980.

Meredith, Jill. "The Arch at Capua: The Strategic Use of *Spolia* and References to the Antique." In Tronzo (1994), 109–126.

Minnis, Alistair J. "Late-Medieval Discussions of *Compilatio* and the Role of the *Compilator.*" *Beitrage zur Geschichte der Deutschen Sprache und Litteratur* 101 (1979): 385–421.

———, ed. *Latin and Vernacular: Studies in Late-Medieval Texts and Manuscripts.* Woodbridge, Suffolk: Boydell and Brewer, 1989.

Minnis, A. J. and A. B. Scott with David Wallace. *Medieval Literary Theory and Criticism c.1100–1375: The Commentary Tradition.* Rev. ed. Oxford: Clarendon Press, 1988.

Mitchell, J. B. "Early Maps of Great Britain I: The Matthew Paris Maps." *Geographical Journal* 81 (1933): 27–34.

Mitchell, Bruce and Fred C. Robinson. *A Guide to Old English.* 4th ed. Oxford: Blackwell, 1986.

Moir, A. L. "*Mappa Mundi*: The World Map in Hereford Cathedral." *Transactions of the*

Caradoc and Severn Valley Field Club 15 (1957–60): 122–127.

———. *The World Map in Hereford Cathedral.* 5th ed. Hereford: Hereford Cathedral, 1970.

Monastic Britain. Map. Ordnance Survey. Southampton, 1978.

Moorman, J. R. H. *Church Life in England in the Thirteenth Century.* Cambridge: Cambridge University Press, 1955.

Morgan, Nigel J. *Early Gothic Manuscripts.* 2 vols. London: Harvey Miller Publishers, Oxford University Press, 1982–88.

Morpurgo, Piero. "*Philosophia naturalis* at the Court of Frederick II: From the Theological Method to the *ratio secundum physicam* in Michael Scot's *De Anima.*" In Tronzo (1994), 241–48.

Morris, J. E. *The Welsh Wars of Edward I.* Oxford: Oxford University Press, 1901.

Morse, Charlotte Cook, Penelope Reed Doob and Marjorie Curry Woods, eds. *The Uses of Manuscripts in Literary Studies: Essays in Memory of Judson Boyce Allen.* Kalamazoo, MI: Medieval Institute Publications, Western Michigan University, 1992.

Murphy, James J. *Rhetoric in the Middle Ages: A History of Rhetorical Theory from Saint Augustine to the Renaissance.* Berkeley: University of California Press, 1974.

Nebenzahl, Kenneth. *Maps of the Holy Land: Images of Terra Sancta through Two Millennia.* New York: Abbeville Press, 1986.

Neubecker, O. *A Guide to Heraldry.* New York: McGraw-Hill, 1979.

Nicolet, Claude. *Space, Geography, and Politics in the Early Roman Empire.* Ann Arbor: University of Michigan Press, 1991.

O'Keefe, Katherine O'Brien. *Visible Song: Transitional Literacy in Old English Verse.* Cambridge: Cambridge University Press, 1990.

Ormrod, W. M., ed. *England in the Thirteenth Century: Proceedings of the 1984 Harlaxton Symposium.* Woodbridge, Suffolk: Boydell Press, 1986.

———, ed. *England in the Fourteenth Century: Proceedings of the 1985 Harlaxton Symposium.* Woodbridge, Suffolk: Boydell Press, 1986.

———, ed. *England in the Thirteenth Century: Proceedings of the 1989 Harlaxton Conference.* Stamford: Paul Watkins, 1991.

Parkes, Malcolm. "The Influence of the Concepts of *Ordinatio* and *Compilatio* on the Development of the Book." In Alexander and Gibson (1976), 115–41.

Parsons, E. J. S. Introduction. *The Map of Great Britain circa AD 1360 Known as the Gough Map.* Oxford: Bodleian Library, 1958.

Partner, Nancy F. *Serious Entertainments: The Writing of History in Twelfth-Century England.* Chicago: University of Chicago Press, 1977.

———. "Making Up Lost Time: Writing on the Writing of History." *Speculum* 61 (1986): 90–117.

Pasquier, Henri. *Un Poeté Latin du xi siecle: Baudri, abbé de Bourgueil.* Paris, 1878.

Patterson, Lee, ed. *Literary Practice and Social Change in Britain, 1380–1530.* Berkeley: University of California Press, 1990.

———. *Chaucer and the Subject of History.* Madison: University of Wisconsin Press,

1991.

Pearsall, Derek. *Old English and Middle English Poetry*. London: Routledge, 1977.

———. *The Life of Geoffrey of Chaucer: A Critical Biography*. Oxford: Blackwell, 1992.

———. "Resident Aliens in Fourteenth-Century London." Conference paper delivered at the University of Minnesota, "Strangers in Medieval Society" Conference. February, 1994.

Pelham, R. A. "Early Maps of Great Britain II: The Gough Map." *The Geographical Journal* 81 (1933): 34–39.

Pelletier, Monique, ed. *Géographie du monde au moyen âge et la Renaissance*. Paris: Editions du C.T.H.S., 1989.

Phillips, J. R. S. "Edward II and the Prophets." In Ormrod (1986b).

———. *The Medieval Expansion of Europe*. Oxford: Oxford University Press, 1988.

Pickles, John. "Texts, Hermeneutics and Propaganda Maps." In Barnes and Duncan (1992), 193–230.

Pirenne, Henri. *Medieval Cities: Their Origins and the Revival of Trade*. Tr. Frank D. Halsey. Princeton, NJ: Princeton University Press, 1969.

Poole, A. L. *From Domesday Book to Magna Carta, 1087–1216*. Oxford: Oxford University Press, 1953.

Powell, James. "Economy and Society in the Kingdom of Sicily under Frederick II: Recent Perspectives." In Tronzo (1994), 268–69.

Powicke, Maurice. *The Thirteenth Century, 1216–1307*. Oxford: Clarendon Press, 1952.

Prestwich, Michael. *War, Politics, and Finance Under Edward I*. London: Faber and Faber Limited, 1972.

———. *The Three Edwards*. London: Weidenfeld Nicolson, 1980.

———. "The Piety of Edward I." In Ormrod (1986a), 120–128.

Putter, Ad. *Sir Gawain and the Green Knight and French Arthurian Romance*. Oxford: Oxford University Press, 1995.

Radke, Gary M. "The Palaces of Frederick II." In Tronzo (1994), 179–88.

Richardson, H. G. "The English Coronation Oath." *Speculum* 24 (1949): 44–75.

Rose, Gillian. *Feminism and Geography: The Limits of Geographical Knowledge*. Minneapolis: University of Minnesota Press, 1993.

Ross, Kristin. *The Emergence of Social Space: Rimbaud and the Paris Commune*. Minneapolis: University of Minnesota Press, 1988.

Rouse, Mary A. and Richard H. Rouse. *Authentic Witnesses: Approaches to Medieval Texts and Manuscripts*. Notre Dame, IN: University of Notre Dame Press, 1991.

Rowlands, Ifor. "The Edwardian Conquest and its Military Consolidation." In Herbert and Jones (1988), 41–72.

Rubin, Miri. *Corpus Christi: The Eucharist in Late Medieval Culture*. Cambridge: Cambridge University Press, 1991.

Said, Edward. *Culture and Imperialism*. New York: Knopf Publishing Group, 1993.

Sanders, Andrew. *The Short Oxford History of English Literature*. Rev. ed. Oxford: Oxford University Press, 1996.

Sanders, W. B. *Map of England and Scotland (author unknown) preserved in the Bodleian*

Library, probable date about AD 1300. Southampton: Bodleian Library, 1875.

Salter, Elizabeth. *England and International: Studies in the Literature, Art, and Patronage of Medieval England.* Cambridge: Cambridge University Press, 1988.

Saul, Nigel. *Knights and Esquires: The Gloucestershire Gentry in the Fourteenth Century.* Oxford: Clarendon Press, 1981.

Sauerlander, Willibald. "Two Glances From the North: The Presence and Absence of Frederick II in the Art of the Empire; The Court Art of Frederick II and the *opus francigenum*." In Tronzo (1994), 189–209.

Scanlon, Larry. "The King's Two Voices: Narrative and Power in Hoccleve's *Regement of Princes*." In Patterson (1990), 216–247.

Scase, Wendy. "'Strange and Wonderful Bills': Bill-casting and Political Discourse in Late Medieval England." *New Medieval Literatures* 2 (1998): 225–248.

Schachterle, Lance and Mark Greenberg, eds. *Literature and Technology.* Lehigh, PA: Lehigh University Press, 1992.

Schulz, Juergen. "Jacopo de' Barbari's View of Venice: Map-Making, City Views and Moralized Geography Before the Year 1500." *Art Bulletin* 60 (1978) 425–74.

Scott, James C. *Domination and the Arts of Resistance: Hidden Transcripts.* New Haven, CT: Yale University Press, 1990.

Sezgin, Fuat. *The Contribution of the Arabic-Islamic Geographers to the Formation of the World Map.* Frankfurt am Main: Johann Wolfgang Goethe University, 1987.

Shichtman, M. R. and J. P. Carley, eds. *Culture and the King: The Social Implications of the Arthurian Legend.* Albany: State University of New York Press, 1994.

Shields, Rob. *Places on the Margin: Alternative Geographies of Modernity.* London: Routledge, 1991.

Sisam, C., ed. *The Vercelli Book: A Late Tenth-Century MS Containing Prose and Verse.* Copenhagen: Rosenkilde and Bagger, 1976.

Skelton, R. A. *Maps: A Historical Study of Their Study and Collecting.* Chicago: University of Chicago Press, 1972.

Skelton, R. A. and E. S. Rafols. "A Contract for World Maps at Barcelona, 1399–1400." *Imago Mundi* 22 (1960): 107ff.

Smith, J. Beverley. "Welsh Society and Native Power before Conquest." In Herbert and Jones (1988), 11–40.

Soja, Edward. *Postmodern Geographies: The Reassertion of Space in Critical Social Theory.* New York: Verso, 1989.

Solterer, Helen. "States of Siege: Violence, Place Gender: Paris around 1400." *New Medieval Literatures* 2 (1998): 95–132.

Soulsby, Ian. *The Towns of Medieval Wales: A Study of their History, Archaeology, and Early Topography.* Chichester, Sussex: Phillimore & Co. Ltd., 1983.

Stein, Robert M. "Making History English: Cultural Identity and Historical Explanation in William of Malmesbury and Layamon's *Brut*." In Tomasch (1998), 97–115.

Stenton, Frank. "The Road System of Medieval England." *Economic History Review* 7 (1936): 7–19.

———. "The Roads of the Gough Map." In Parsons (1958), 16–20.

Stevens, C. E. "The *Notitia Dignitatum* in England." In Goodburn and Bartholomew (1976), 211–224.

Stevenson, B. *The Home Book of Proverbs, Maxims and Familiar Phrases.* New York: MacMillan, 1948.

Stevenson, Edward Luther. *Portolan Charts: Their Origin and Characteristics with a Descriptive List of those belonging to the Hispanic Society of America.* New York: Knickerbocker Press, 1911.

Stock, Brian. *The Implications of Literacy: Written Language and Models of Interpretation in the Eleventh and Twelfth Centuries.* Princeton: Princeton University Press, 1983.

Stones, E. L. G. and Grant G. Simpson. *Edward I and the Throne of Scotland, 1290–1296.* Vol. I: Introduction. Oxford: Oxford University Press, 1978.

Strayer, J. R. "The Laicization of French and English Society in the Thirteenth Century." *Speculum* 15 (1940): 76–86.

———, ed. *The Dictionary of the Middle Ages.* 13 vols. New York: Charles Scribner's Sons, 1983–85.

Strohm, Paul. *Hochon's Arrow: The Social Imagination of Fourteenth-Century Texts.* Princeton, NJ: Princeton University Press, 1992.

Tasker, E. and J. Beaumont. *Encyclopedia of Medieval Church Art.* London: Batsford, 1993.

Tatlock, J. S. P. *The Legendary History of Britain: Geoffrey of Monmouth's Historia Regum Britanniae and its Early Vernacular Versions.* Berkeley: University of California Press, 1950.

Tattersall, Jill. "Sphere or Disc? Allusions to the Shape of the Earth in Some Twelfth-Century and Thirteenth-Century Vernacular French Works." *Modern Language Review* 76 (1981) 31–46.

Taylor, E. G. R. *Tudor Geography, 1485–1583.* London: Methuen & Co., 1930.

———. "The Hereford Map in Facsimile." *Geographical Journal* 120 (1954): 221–23.

Thrower, N. J. *Maps and Man: An Examination of Cartography in Relation to Culture and Civilization.* Englewood Cliffs, NJ: Prentice Hall, 1972.

Tomasch, Sylvia. "*Mappae Mundi* and 'The Knight's Tale': The Geography of Power, the Technology of Control." In Greenberg and Schachterle (1992), 66–98.

——— and Sealy Gilles, eds. *Text and Territory: Geographical Imagination in the European Middle Ages.* Philadelphia: University of Pennsylvania Press, 1998.

Tomlin, R. S. O. "Notitia Dignitatum Omnium, Tam Civilium Quam Militarium." In Goodburn and Bartholomew (1976), 189–209.

Tooley, R. V. *Maps and Map-Makers.* London: B. T. Batsford, 1949.

Tristam, E. W. *English Medieval Wall-Painting: The Thirteenth Century.* Oxford: Oxford University Press, 1950.

Tronzo, William, ed. *Intellectual Life at the Court of Frederick II.* Washington, DC: National Gallery of Art, 1994.

Tudor-Craig, Pamela. "The Painted Chamber at Westminster." *Archaeological Journal* 114 (1957): 90–105.

Vaughan, Richard. *Matthew Paris.* Cambridge: Cambridge University Press, 1958.

———. *The Illustrated Chronicles of Matthew Paris: Observations of Thirteenth-Century Life*. Cambridge: Alan Sutton Publishing, 1993.

Vermette, Rosalie. "Terrae Incantatae: The Symbolic Geography of Twelfth-Century Arthurian Romance." In *Geography and Literature: A Meeting of the Disciplines,* ed. William E. Mallory and Paul Simpson-Housley, 145–60. Syracuse: Syracuse University Press, 1987.

von den Brincken, Anna-Dorothee. "Monumental Legends on Medieval Manuscript Maps." *Imago Mundi* 42 (1990).

Waddell, Helen. *The Wandering Scholars*. Boston: Houghton Mifflin Co., 1927.

Wall, J. Charles. *Shrines of British Saints*. London: Methuen & Co., 1905.

Wallace, David. "Chaucer and the Absent City." In Hanawalt (1992), 59–90.

———. *Chaucerian Polity: Absolutist Lineages and Associational Forms in England and Italy*. Stanford: Stanford University Press, 1997.

———, ed. *The Cambridge History of Medieval English Literature*. Cambridge: Cambridge University Press, 1999.

Wallis, Helen, ed. *Historians' Guide to Early British Maps*. London: David Brown Book Co., 1994.

——— and Sarah Tyacke, eds. *My Head Is a Map: Essays and Memoirs in Honor of R. V. Tooley*. London: Francis Edwards and Carta Press, 1973.

Watkins, Alfred. *Early British Trackways, Moats, Mounds, Camps, and Sites*. Hereford: Watkins Meter Co., 1922.

Westrem, Scott. "Against Gog and Magog." In Tomasch and Gilles (1998), 54–75.

Whitfield, Peter. *The Image of the World: Twenty Centuries of World Maps*. San Francisco: Pomegranate Art Books, 1994.

Williams, Daniel. "Matthew Paris and the Thirteenth Century Prospect of Asia." In Ormrod (1991), 51–67.

Wiseman, Peter. "Julius Caesar and the Hereford World Map." *History Today* 37 (1987): 53–57.

Wittkower, Rudolph. "Marvels of the East: A Study in the History of Monsters." *Journal of the Warburg and Courtauld Institutes* 5 (1942): 159–197.

Wolf, Armin. "News on the Ebsdorf World Map: Date, Origin and Authorship." In Pelletier (1989), 51–68.

Wogan-Browne, J. "Reading the World: the Hereford *Mappa Mundi*." *Parergon: Bulletin of the Australian and New Zealand Association of Medieval and Renaissance Studies* 9.1 (June 1991): 117–135.

Woodward, David. "Reality, Symbolism, Time, and Space in Medieval World Maps." *Annals of the Association of American Geographers* 75 (1985): 510–21.

———. "Medieval *Mappaemundi*." In Harley and Woodward (1987) 286–370.

———, ed. *Art and Cartography: Six Historical Essays*. Chicago: University of Chicago Press, 1987.

———. "Medieval World Maps." In Pelletier (1989), 1–8.

Woodward, J. and G. Burnett. *A Treatise on Heraldry*. Rutland, VT: C.E. Tuttle, 1969 [1896].

Wormald, Francis. *Collected Writings, vol. II: Studies in English and Continental Art.* London: Harvey Miller, 1988.

Wright, John Kirtland. *The Geographical Lore of the Time of the Crusades: A Study in the History of Medieval Science and Tradition in Western Europe.* New York: American Geographical Society, 1925.

Yates, W. N. "The Authorship of the Hereford *Mappa Mundi* and the Career of Richard de Bello." *Transactions of the Woolhope Naturalists' Field Club* 41 (1974): 165–72.

Index of Names, Places, and Subjects Mentioned in Text

[Please note: Material appearing in Notes (pp. 163-213) has not been included; also omitted have been some names and places mentioned in the text itself—typically those appearing singly or incidentally; medieval names have been alphabetized by first name or surname variously, according to standard usage ("Geoffrey of Monmouth" but "Mannyng, Robert, of Bourne"); underlined page-numbers indicate an illustration or figure-caption.]

A

F

H

I

N

R

S

T

Y

Z